RA 569 .R25 1995

Radiation dose
reconstruction for

DATE DUE

RADIATION DOSE RECONSTRUCTION

for

Epidemiologic

Uses

Committee on an Assessment of CDC Radiation Studies
Board on Radiation Effects Research
Commission on Life Sciences
National Research Council

NATIONAL ACADEMY PRESS
Washington, D.C. 1995

NATIONAL ACADEMY PRESS 2101 CONSTITUTION AVENUE, N.W., WASHINGTON, D.C. 20418

The project that is the subject of this report was approved by the governing board of the National Research Council, whose members are drawn from the councils of the National Academy of Sciences, the National Academy of Engineering, and the Institute of Medicine. The members of the committee responsible for the report were chosen for their special competences and with regard for appropriate balance.

This report has been reviewed by a group other than the authors according to procedures approved by a Report Review Committee consisting of members of the National Academy of Sciences, the National Academy of Engineering, and the Institute of Medicine.

This project was prepared under contract 200-91-0951 between the National Academy of Sciences and the Centers for Disease Control and Prevention.

Library of Congress Cataloging-in-Publication Data

Radiation dose reconstruction for epidemiologic uses / Committee on an
    Assessment of CDC Radiation Studies, Board on Radiation Effects
    Research, Commisson on Life Sciences, National Research Council.
        p.   cm.
    Includes bibliographical references and index.
    ISBN 0-309-05099-5
    1. Radiation injuries—Epidemiology—Statistical Methods.
    2. Radiation dosimetry. I. National Research Council (U.S.).
    Committee on an Assessment of CDC Radiation Studies.
    RA569.R25   1995
    616.9'897—dc20   95-10524

The National Academy of Sciences is a private, nonprofit, self-perpetuating society of distinguished scholars engaged in scientific and engineering research, dedicated to the furtherance of science and technology and to their use for the general welfare. Upon the authority of the charter granted to it by the Congress in 1863, the Academy has a mandate that requires it to advise the federal government on scientific and technical matters. Dr. Bruce M. Alberts is president of the National Academy of Sciences.

The National Academy of Engineering was established in 1964, under the charter of the National Academy of Sciences, as a parallel organization of outstanding engineers. It is autonomous in its administration and in the selection of its members, sharing with the National Academy of Sciences the responsibility for advising the federal government. The National Academy of Engineering also sponsors engineering programs aimed at meeting national needs, encourages education and research, and recognizes the superior achievements of engineers. Dr. Robert M. White is president of the National Academy of Engineering.

The Institute of Medicine was established in 1970 by the National Academy of Sciences to secure the services of eminent members of appropriate professions in the examination of policy matters pertaining to the health of the public. The Institute acts under the responsibility given to the National Academy of Sciences by its congressional charter to be an adviser to the federal government and, upon its own initiative, to identify issues of medical care, research, and education. Dr. Kenneth I. Shine is president of the Institute of Medicine.

The National Research Council was organized by the National Academy of Sciences in 1916 to associate the broad community of science and technology with the Academy's purposes of furthering knowledge and advising the federal government. Functioning in accordance with general policies determined by the Academy, the Council has become the principal operating agency of both the National Academy of Sciences and the National Academy of Engineering in providing services to the government, the public, and the scientific and engineering communities. The Council is administered jointly by both Academies and the Institute of Medicine. Dr. Bruce M. Alberts and Dr. Robert M. White are chairman and vice chairman, respectively, of the National Research Council.

Preface

As PUBLIC CONCERN MOUNTS over past and current exposure to ionizing radiation stemming from environmental releases of radioactive materials, there is a growing need to define the criteria to be met by studies that reconstruct exposures and doses and to provide guidance in the studies' epidemiologic use. Absent this, dose reconstruction studies are not likely to stand serious scientific scrutiny or to meet public concerns. To assist the Centers for Disease Control and Prevention (CDC) in the continuing dose reconstruction efforts at several U.S. nuclear facilities, the members of the National Research Council's Committee on an Assessment of CDC Radiation Studies and officers at the CDC believed it was timely to convene a group of scientists with experience and expertise relevant to the dozens of major dose reconstruction projects around the world that have followed radiation exposures of human populations. The scientists were asked to assist the committee in identifying criteria to be considered when undertaking radiation dose reconstruction studies, to examine the pitfalls encountered in previous studies, and to recommend areas of needed research. This report should set the objectives to be attained by such studies and provide guidelines for their conduct. It is aimed at providing generic information to scientists entering the field and to interested members of the public.

The National Research Council committee is indebted to the numerous scientists from around the world who agreed to participate in the Workshop on Dose Reconstruction for Epidemiologic Uses, which was held in Washington, D.C., Oct. 25-27, 1993. The document that follows was synthesized by these scientists who, working with the committee, put into writing their thoughts and experiences. This final document was edited by the National Research Council committee and was subjected to the Research Council's rigorous and independent review process. It is important to emphasize that this version has not been reviewed by all of the participants and that it does not claim to represent a consensus of the workshop participants. The committee realizes that it would be impractical to achieve such extensive review and consensus, given the large number of participants. However, the committee believes this document captures the enthusiasm and conscientiousness displayed by the participants. It is a reasonably accurate record of their thoughts, it defines a valuable set of criteria, and it provides recommendations that will prove useful in future dose reconstruction studies in the United States and elsewhere.

We are deeply appreciative of the work of the staff of the Board on Radiation Effects Research, and particularly the assistance of Doris Taylor and Maurita Dow-Massey in the preparation of this report. We thank Mrs. Kate Kelly for her editorial review.

WILLIAM J. SCHULL
Chairman

Contents

RADIATION DOSE RECONSTRUCTION

for

Epidemiologic

Uses

Executive Summary

AT THE OUTSET OF THE nuclear age, and for some years thereafter, radioactive materials were released to the environment at many places. Because these materials might have affected the health of populations living near the sites of the releases, public concern has led to the need to reconstruct doses, to define how such reconstructions should be accomplished, and to determine how the results should be used. Dose reconstructions must meet two criteria: They must withstand scientific scrutiny, and they must satisfy public concern. The Committee on an Assessment of CDC Radiation Studies was convened to provide scientific advice to the National Center for Environmental Health and Injury Control of the Centers for Disease Control (CDC) and Prevention and to evaluate the quality and completeness of CDC's dose reconstructions and epidemiologic studies. To assist with continuing dose reconstruction studies and to help in planning those that might be conducted in the future, a workshop was convened to identify criteria for dose reconstruction studies and to recommend needed areas of research. Forty-seven scientists from around the world with expertise in dose reconstruction, dose and risk assessment, and epidemiology, among other disciplines, were invited to the workshop, which summarized the state of the science, highlighted lessons of the past, and delineated general principles useful for undertaking radiation dose reconstruction studies. These principles could be extended to chemical hazards.

Discussions during the workshop and subsequent reviews have led the committee to recommend that preliminary studies, to be called scoping studies, be performed before full-scale dose reconstruction studies begin. Scoping studies should establish the magnitude of the estimated exposure, the size and the composition of the potentially affected population, and the feasibility of conducting rigorous dosimetric and epidemiologic studies. The committee made two recommendations:

• Scoping studies should be based on realistic assumptions about the dose distribution, the population size, and the expected harm to public health resulting from a radiation release.

• Full-fledged dose reconstruction and epidemiologic studies should be proposed only when scoping studies show that rigorous studies are possible, given the probable dose distribution and size of the exposed population.

Full-scale dose reconstruction and epidemiologic studies should provide for the following elements:

• All pertinent data relating to the source term and environmental pathways should be collected and evaluated. Insofar as possible, the original source data, rather than derived or summary information, should be used.

• Quality control should be required at all stages of data collection and evaluation. Where possible, alternative approaches should be used to estimate the components of the dosimetry (source term, environmental transport, metabolic disposition, behavioral variation).

• Best estimates of doses should be used rather than maximal doses; uncertainties in doses (defined by confidence intervals) also should be estimated.

• Biologic markers of dose, effect, and susceptibility should be considered. The committee cautions, however, that at most of the sites of potential concern in the United States, the doses are probably not large enough for biologic markers to be useful.

The committee recommends that scoping of both the dosimetry and the epidemiology be performed interactively and in parallel because both are needed to inform decisions about further studies of one site or to establish priorities among several. This also applies to full-scale studies because it is important to have epidemiologists involved from the outset of any dose reconstruction to ensure that dosimetric information is appropriate for epidemiologic use.

The committee recommends that questions that are to be answered by any dose reconstruction study should be specified carefully and that the scientists involved should work with persons who have an interest in the study. The committee further recommends that studies be conducted with full disclosure and with the greatest degree of public participation practical. There should be an interactive process that communicates the details and methods of the study in a timely and understandable manner to the public, and an oversight committee, consisting of impartial scientists and citizens whose meetings are open to the public, should be established at the earliest possible moment and maintained throughout the study.

Finally, the committee recommends that, to arrive at a credible and cost-effective decision-making process for identifying and prioritizing sites for study, the criteria used to proceed from a scoping study to a comprehensive dose reconstruction study should be adopted before the scoping study begins.

1

Background

U NDER A MEMORANDUM of under-
standing concluded with the U.S. Department of Energy (DOE) in December 1990, the Department of Health and Human Services, through its
Centers for Disease Control and Prevention (CDC), has undertaken a
series of studies to assess the possible health consequences of off-site
emissions of radioactive materials from DOE-managed nuclear facilities
in the United States. At the request of CDC, the Board on Radiation
Effects Research, in the National Research Council's Commission on Life
Sciences, has organized the Committee on an Assessment of CDC Radiation Studies to provide scientific advice to CDC's National Center for
Environmental Health and Injury Control. The committee's charge was
as follows:

• Review and comment on the design, methods, analysis, statistical
reliability, and scientific interpretation of dose reconstruction and related
epidemiologic follow-up studies.
• Recommend ways to strengthen study protocols and analyses to
enhance the quality of these studies.

In the course of the committee's review of CDC's dose reconstruction
efforts in the vicinity of the Fernald, Ohio, Feed Materials Production
Center (NRC 1992, 1994a) and the Hanford Nuclear Site in southeast
Washington State (NRC 1994b, 1995a), the committee and the National
Center for Environmental Health and Injury Control decided that it would
be timely to assemble scientists with expertise in the major areas of radia-

tion dose reconstruction. The goal was to define the criteria to be met in studies that reconstruct exposure and to provide guidance for their epidemiologic use.

Radiation dose reconstruction methodologies are developing rapidly. Dose reconstruction projects during the past 50 years have resulted in an accumulation of considerable experience. Such projects have been considered at Hanford, Washington; Fernald, Ohio; and the Sellafield Nuclear Processing Facility, United Kingdom. Dose reconstruction also has been suggested in the aftermath of accidents in such places as Chernobyl, Ukraine; Goiania, Brazil; Palomares, Spain; and Kyshtym, Russia; and after the detonation of nuclear weapons (Hiroshima and Nagasaki, Japan, and at the Nevada and Pacific test sites). Little effort has been made to assemble the lessons learned from these projects or to identify criteria for a cost-effective dose reconstruction. The committee sought a remedy through a workshop designed to achieve three objectives:

• Summarize past and current dose reconstruction studies in the United States and elsewhere, detailing the techniques used and the scientific problems encountered.

• Establish not only the criteria for a thorough dose reconstruction but minimum requirements for a study that is used to determine possible health effects.

• Identify the information needed to specify doses for particular persons or groups to be used in epidemiologic studies.

A 3-day workshop was held at the National Academy of Sciences in Washington, D.C., Oct. 25-27, 1993, which convened 47 scientists from around the world with experience in one or more of the four areas listed below that are central to the reconstruction of radiation doses resulting from releases of radioactivity:

• Estimating the environmental release of radioactive material (the so-called source term).

• Environmental pathway analysis, which leads to estimates of radionuclide deposition on the ground or in surface and groundwater and to estimates of radionuclide concentrations in ground-level air, drinking water, and foodstuffs.

• Identifying exposed populations and collecting dietary and lifestyle data to facilitate valid estimates of exposure to external irradiation and of the inhalation and ingestion of radionuclides.

• Dose assessment for specified persons or population groups or for representative individuals in a general population.

The degree to which any of these areas can be studied effectively varies from one site or episode to another because the circumstances of each event differ, for example, in the meteorologic conditions during the episode, in the nature of the terrain, and in the sources of surface water and groundwater. However, the participants were asked to identify a set of general criteria that would apply to most dose reconstructions. To do this, participants were assigned to one of five focus groups to discuss and define the important steps of dose reconstruction. After they attended a series of general lectures in each area, the group members were charged with summarizing knowledge in their topic areas, highlighting the lessons of the past, and identifying new areas for research. The summaries drafted by the groups form the bases of the chapters in this report. Chapters 3-7, respectively, are "Estimating and Confirming the Source Term," "Environmental Pathways," "Radiation Dose Assessment," "Biologic Dosimetry and Biologic Markers," and "Epidemiologic Considerations."

During the workshop, the need for a sixth working group was identified to provide guidelines for priority criteria for dose assessment studies. Chapter 8 summarizes that group's deliberations and addresses the process of setting priorities for dose assessment studies that use criteria based on scientific evidence. Appendix A briefly describes seven dose reconstruction studies, Appendixes B and C are the workshop agenda and a list of participants, respectively, and Appendix D is a glossary.

2

Introduction

OVER THE PAST HALF CENTURY, the United States government has built many facilities for the design and construction of nuclear weapons and for processing radioactive materials. Particularly in the early years of operation, the radioactive discharges from some of these installations could have been high enough to pose a potential hazard to nearby residents. Concern over the possible exposure has prompted efforts to reconstruct doses because the potential hazard can be evaluated only through documentation of the dose.

In this document, dose reconstruction is defined as the process of estimating doses to the public from past releases to the environment of radionuclides or chemicals. These doses form the basis for estimating health risks and for determining whether epidemiologic studies are warranted. Past exposures, not current ones, are the focus of this report. It is important to recognize that not all releases have led to public exposure. Moreover, in some cases, the releases have already occurred, but exposure will happen only in the future—for example, in the case of contaminated groundwater that has not migrated off-site. The methods used to estimate risk from a hypothetical release and to estimate the resulting rate and pathway of off-site transport of contaminants can be quite different from those used for retrospective assessments.

Although terms such as "low" and "high" are used often to describe exposures or doses, they are subjective, and when they are applied to doses or dose rates they offer little quantitative guidance. To avoid ambiguity, the committee uses "low" and "high" to describe the magnitude of the dose; the terms do not connote a judgment about the significance of the radiation exposure. The definitions recommended by the United Nations Scientific Committee on the Effects of Atomic Radiation (UNSCEAR 1993; see pp. 680-682 and Table 8, Annex F) are used here: A dose is "low" if it does not exceed 0.2 gray[1] (Gy, the international unit for absorbed dose) or 20 rad (rad is a unit of absorbed dose), whatever the dose rate. Similarly, a dose rate is "low" if it does not exceed 0.006 Gy hr⁻¹ (6 mGy hr⁻¹ or 0.6 rad hr⁻¹), whatever the accumulated dose. A "high" dose rate, at least in experimental studies, usually is in the range of 6 to 48 Gy hr⁻¹ (600 to 4,800 rad hr⁻¹). Doses above 0.2 Gy (20 rad) generally are subdivided into those that are "intermediate," 0.2-2.0 Gy (20-200 rad), and those that are "high," >2 Gy (>200 rad). As a point of reference, the lifetime (70-year) dose from natural background and medical radiation for the average U.S. resident is about 0.3 Gy (30 rad).

It is important to note that serious health effects of exposure to ionizing radiation, such as an increase in cancer, have not been observed directly at doses below 0.2 Gy (20 rad) among the survivors of the atomic bombing of Hiroshima and Nagasaki. The risks assumed to occur at doses below 0.2 Gy (20 rad) are, therefore, extrapolations from the risks seen at intermediate and high doses to doses above natural background radiation.

ELEMENTS OF DOSE RECONSTRUCTION

Dose reconstruction studies typically strive to estimate representative doses, doses to specific persons, or both. Representative doses are doses to people who have received an average dose in a particular region and not doses to specific individuals who lived near a site. Generally, they can be derived from historic records or from data from the scientific literature that describe diet, lifestyle, and history for people in typical age categories in the area. Representative doses illustrate the magnitude of the dose and the importance of specific pathways and contaminants in general situations. Such doses can be used to determine the potential statistical power of a proposed epidemiologic study. Individual doses are estimated for specific real people, and the studies use demographic, residential, or dietary data that these persons provide for the dose calculation. Depending on the circumstances and types of exposure, an epidemiologic study can use individual doses or representative doses. However, epidemiologic studies that use individual doses generally are considered more accurate than are studies that use representative doses. Individual doses also best address public concerns.

The criteria for the design of a dose reconstruction project must be expressed in terms of specific questions. These criteria differ depending on whether the project is intended primarily to answer questions posed by the public or by researchers who want to learn more about the health effects of radiation. Statistically detectable effects that are of great scientific interest might have little public health significance. That is, a particular health effect can be too small or uncommon to warrant any change in public health policy (for example, offering cancer screening to all the residents of a given area). Moreover, it is probably better public health policy to use limited public health resources to prevent future exposures to radiation (for example, with a readily available home radon abatement program) than it is to detect and treat the consequences of past exposures.

TECHNICAL ASPECTS OF DOSE RECONSTRUCTION

The analysis of historic data from a site or operation to determine off-site radiation doses involves several steps:

• *Source term analysis* consists of estimating the magnitude of releases to the environment of radionuclides and the periods over which they were released, including episodic releases from nonroutine events.

• *Pathway analysis* examines the transport of released radionuclides through environmental pathways to determine their concentrations in environmental media to which people were exposed. These media include air, surface and groundwater, and soil, among others.

• *Assessment of radiation doses and risks* brings together all of the data on releases, transport, lifestyle and dietary habits, analysis of agricultural and food-distribution practices, and biologic factors, including the use of biologic dosimetry, to determine doses or to corroborate evidence of doses and to estimate the likelihood of disease in the exposed persons.

• *Examination of epidemiologic considerations* takes into account the size and demographic structure of the potentially affected population, the availability and quality of information needed to estimate the dose, the medical information needed, and the feasibility of conducting an investigation that is sufficiently informative and free of bias.

• *Uncertainty and sensitivity analysis* identifies the importance of changes in the parameters and values used to estimate confidence intervals in the overall analysis of the dose reconstruction (see Glossary for distinction between uncertainty and sensitivity analysis).

As mentioned previously, a limited number of dose reconstruction studies have been done and there are lessons to be learned from these efforts. Seven studies, their approaches to dose estimation, and their pitfalls are summarized in Appendix A.

Often, as the studies set out in Appendix A attest, a simple approach, which we term a "scoping study," can determine whether a more comprehensive dose reconstruction study is warranted or even possible, either for its own purpose or as the basis for an epidemiologic study. If a full dose reconstruction study is justified, all relevant processes should be identified and quantified for the dose-determining radionuclides emitted from the source or the manufacturing practice under study. The source term characteristics (the chemical and physical form, time dependence, etc., of the release) and the specific geographic, agricultural, and meteorologic conditions associated with the site, also should be determined.

Scoping estimates can be made with bounding calculations and relatively simple models. They are intended merely to provide guidance about the amount of investigation required for subsequent stages in the dose reconstruction study. Scoping estimates are difficult to make, however, unless there is an established set of starting and stopping rules. The stopping rules could be based on exposure levels associated with epidemiologic feasibility, exposure levels that are commensurate with health risk, or exposure levels that are clearly a negligibly small fraction of the estimated total exposure. Scoping estimates are most useful for identifying episodes that require more detailed investigation. Scoping studies are discussed further in the following section.

Historic records are commonly the foundation of a dose reconstruction project, and it is always preferable to use measured data (historic records) rather than models in the reconstruction of doses. The use of previous summaries of data predicting radiation exposures or environmental concentrations should be avoided if possible, especially for the years being reconstructed. Instead, emphasis should be given to the basic data, such as the records made by workers involved in actual operations and to quality control of the data. If it can be shown that the basic data consistently match the information given in summary reports, then the use of summary documents can be defended (Till 1993).

There are two common types of searches of the historic records. In a selective search, documents are sought that are clearly needed to support the research, and specific pieces of information are sought as the study progresses. Its advantage is that finite resources are saved by not having to review all or even a large fraction of the accumulated records. The disadvantage is that important data can be overlooked, possibly leading to deficiencies in the dose reconstruction, and conceivably to a loss of credibility (Till 1993). In a comprehensive search, all records of potential importance are reviewed and catalogued before the dose reconstruction starts. This approach is likely to strengthen the dose reconstruction effort. Whichever approach is used, it is important that the representativeness of the data that are used in the dose reconstruction is carefully evaluated.

Epidemiologic study of populations requires the use of quantitative dose reconstruction data. Because different levels of epidemiologic investigation demand different amounts of detail and precision in dose estimates, dosimetry and epidemiologic screening efforts will be most informative if they are conducted interactively and in parallel. Delaying an epidemiologic investigation until the dose reconstruction is complete could diminish the usefulness of the reconstruction and jeopardize the epidemiologic study. For example, the farther one gets from the time period of interest the more difficult it becomes to assemble the complete population and to reconstruct the information needed for exposure assessment. Moreover, the likelihood of a variety of biases increases as the publicity surrounding a dose reconstruction broadens. Finally, negative public perception brought about by delays in directly addressing the issues of concern could require additional efforts that cannot be justified scientifically.

STRUCTURE OF A SCOPING STUDY

Several steps can be identified in the conduct of a scoping study as outlined in Figure 2-1. The call for a study often begins with public input or an expression of strong public concern about observed or suspected health effects experienced by the population near a site. Any data and evidence that supports public concerns are evaluated, and the accessible data are identified. The feasibility and plausibility of a potential study are assessed based on evidence of probable effects. A preliminary epidemiologic assessment is made to estimate the affected population size, identify the demographic composition of the population, review available medical data, estimate the study's statistical power, and assess the ability to conduct a rigorous study in which biases are minimized. A parallel, preliminary dose assessment is done to describe major release streams and environmental pathways, the nature of effluents of concern, and the target population and to give a best estimate of annual organ doses to representative exposed individuals. A qualitative severity assessment, based on conservative and nonconservative ("realistic") assumptions, is done separately for the dosimetric and epidemiologic studies. These could be high-versus-low assessments or they could be based on a point rating scale to be compared against the decision criteria discussed below. Finally, a site is given a priority ranking by comparison with information from similar scoping studies conducted for other sites using the same rules. This process is necessarily iterative, and it is subject to the emergence of additional information including that resulting from any ongoing public involvement.

To summarize briefly,

- For dose reconstruction, a scoping study provides preliminary

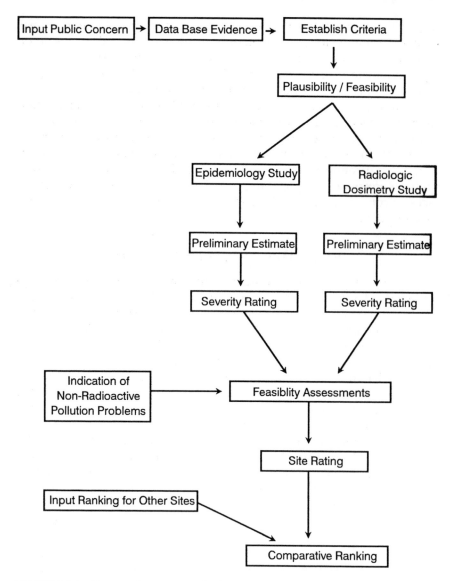

FIGURE 2-1 Proposed structure of a dose reconstruction scoping study.

estimates of source term (if needed), environmental transport (if needed), radiation and radionuclide exposures, and radiation dose. A comprehensive, detailed study provides more refined estimates of the same quantities.

  • For epidemiology, the first stage is more of a feasibility study—it identifies the population at risk and the available data and predicts the sampling sizes needed to proceed with a comprehensive, detailed study to look for health effects and relationships between health effects and dose.

PUBLIC INVOLVEMENT

If dose reconstruction studies are to receive public acceptance, members of the public *must* be involved in their design and implementation. In the past, however, dose reconstruction and epidemiologic studies often were conceived and conducted by individual scientists or by groups of scientists pursuing new information in relative isolation from the lay community. The results of such studies were nevertheless viewed as credible. Times have changed. The public demands that exposed populations be involved in the development and conduct of dose reconstruction studies. The studies frequently are conducted in an atmosphere of suspicion and distrust, with many in the study community believing that their concerns are inadequately addressed. This need to involve concerned parties from the beginning planning and throughout the whole process when risk assessment is involved was recently emphasized by another National Research Council Committee (NRC 1994c).

There are at least three important reasons for public involvement: First, public involvement will help ensure that the public's concerns will be addressed; second, it is the only way the public will have confidence in the results of the study; and third, members of the public sometimes have information that is otherwise unavailable to the investigators. For example, local citizens might be aware of data sources, historic documents, or local traditions unknown to the investigators.

As soon as possible after a decision is made to study a contaminated area, an advisory or steering committee—an oversight and decision-making body—should be constituted. Members of the public should be appointed, should have access to all information, and should participate equally in decision-making. The deliberations of this advisory committee should be public—the public must be involved through its representatives on the advisory committee. Researchers must make regular progress reports and seek advice on public concerns, making certain that the ideas of the public are considered seriously. Although the public must be involved in oversight and decision-making, the scientific aspects of the study should remain the responsibility of the scientists.

To maximize public participation, a public announcement of the proposed study should be made and there should be an open meeting at which the need for and conduct of the study are explained and an opportunity for questions is provided. Public meetings are only one means of engaging the public and enlisting its involvement. Electronic and print media should also be used. Members of the press can be allies, not adversaries, in informing the public. Other educational efforts in schools, churches, and service organizations should be considered and undertaken where appropriate. Workshops, which are less formal than public meetings, have proved especially useful in the Fernald Dosimetry Reconstruction Project, permitting the public and the investigators to focus on specific topics and to discuss the scientific aspects of the study. A reading room where relevant materials are gathered is an excellent means of allowing the public to become better informed.

In addition to public participation, other means exist for ensuring the credibility of the study. One of these is through periodic review of the study by scientists who are not engaged in its conduct and have no interest, or appearance of interest, in its outcome.

In some communities, citizen groups have organized ad hoc epidemiologic studies. Such studies should be discouraged unless they are based on sound scientific principles.

After a dose reconstruction study is completed, particular attention should be given to how the results of the study are to be announced. Preliminary analyses or tabulations are often misinterpreted. The procedure for public release of the results should be made clear at the beginning, and whatever that procedure, the results should be presented in ways that will maximize public understanding. The reconstructed external exposures and exposures to ingested or inhaled radionuclides should be given as ranges rather than as point estimates only, and the meaning of such ranges should be explained clearly by comparing the exposures to common exposures such as natural background or common medical radio-diagnostic procedures.

SUMMARY AND RECOMMENDATIONS

Dose reconstruction studies must rely on solid science, state-of-the-art methods, and careful peer review if they are to be viewed as credible. Ultimately, a dose reconstruction study will be judged by the scientific community primarily on the basis of the technical quality of the study and its contribution to science. However, what the public seeks from a dose reconstruction project is accuracy and candor. Openness, early and continuous public involvement, and clear communication of the study's findings as it progresses can improve the scientific quality of the work. Meaningful public involvement is essential to the success of dose reconstruction studies. To achieve these ends, the committee makes seven general recommendations focusing on the organizational aspects of all dose reconstruction studies:

1. In dose reconstruction studies, thorough consideration should be given to the collection of representative data, to an evaluation of their representativeness, to quality control, and to public involvement from the time the decision-making process begins, through the release of any results, and after the study concludes in any follow-up activities.

2. An advisory or steering committee should be established at the outset of a dose reconstruction study. This committee should consist of members of the public and knowledgeable scientists who are not associated with the investigators or their sponsors. The meetings of this committee should be open to the public. Along with its steering duties, this committee should be charged with responsibility for establishing an interactive process to communicate the elements and conclusions of the study to the general public.

3. Dose reconstruction studies should begin with a scoping study—a preliminary analysis—to determine whether a comprehensive dose reconstruction study is needed or even possible, either for its own purposes or as the basis for a comprehensive epidemiologic study.

4. All dose reconstruction studies should be reviewed by groups of scientists and public health officials who are not directly involved in the study, either as participants or as advisors, and time and resources should be allocated for resolving discrepancies in the results.

5. There should be coordination between the dosimetric and epidemiologic efforts, which should begin at the outset of the dosimetric study and continue throughout.

6. Premature dissemination of the results should be avoided. Results should not be disclosed until the dose assessment study is complete, has undergone peer review, and has been published. Dissemination of data being considered during the study is appropirate and desirable.

7. A clear understanding of the public's concern should be gained before the study begins. A dose reconstruction study should not proceed until the design is such that it is likely that the results will address the public's concern.

NOTE

1. In this report, units related to ionizing radiation will be given as international units (International System of Units, SI) derived from the seven defined units adopted in 1960 at the eleventh Conférénce Générale des Poids et Mesures (Bureau International des Poids et Mesures 1991). An account of SI units and their application can be found in NCRP Report No. 82 (NCRP 1985). However, values expressed using the SI units of absorbed dose (gray), dose equivalent (sievert), and activity (becquerel) will be followed in parentheses by values expressed using the more traditional units of rad, rem, and curie, respectively. The prefix milli (m) represents one-thousandth, or 0.001, and micro (μ) represents one-millionth, or 10^{-6}.

3

Estimating and Confirming the Source Term

T HE SOURCE TERM IS THE amount of radionuclides released from a site to the environment over a specific period. The rate of release as a function of time should also be determined. Releases can be to the atmosphere, to surface waters, to groundwater, or to soil. In some cases, people might have been exposed directly.

A full description of the source term includes what was released and in what form and where and when the release occurred. These factors must be described with enough accuracy and detail both to satisfy the scientific requirements for the design and conduct of epidemiologic studies and to address the public's concern. Satisfying public concern can be difficult given the inherent uncertainty of retrospective dose reconstruction studies. In general, scientific requirements are most likely to be satisfied when it is possible to estimate the source term by several different methods. Usually, a scoping study should precede a full description of the source term; the objective of a scoping study is to provide a preliminary assessment of the source term and the magnitude of the estimated exposure to be used in further decision-making. Ideally, the results of a comprehensive source term analysis will contain all of the information needed for exposure or dose calculations, with a spatial and time resolution sufficient for the requirements of the epidemiologic study.

This chapter is concerned with general approaches to estimating releases rather than with the development of specific source terms. The following sections discuss locations and characteristics of releases; information needed for source term estimates and where that information can be found; biases and uncertainties in historic measurements; and the uses of interpolation, extrapolation, and modeling for estimating source terms.

APPROACH TO SOURCE TERM ANALYSIS

There are, in principle, three ways to determine the source term in the case of a specific facility—from engineering estimates of what the operation was expected to release, from historic reports of measured releases, and from reconstruction that uses independent measurements of environmental quantities related to the release. Not all of these ways will be possible in all cases, but to the extent they are, redundant analyses are desirable. A source term analysis consists of a mix of information about the physical, engineering, and chemical elements of a source. The information comes from models or from historic records provided by production data, from environmental monitoring of the amounts and forms of released materials and the environmental pathways into which the material was introduced (air, water, land), and from contemporary measurements made in the environment at the time of the dose reconstruction.

The quality and completeness of the estimates of source terms, based on the physical and chemical characteristics of the source supplemented by historic, operational, and release data and those based on any available independent environmental measurements, vary with the facility or site being evaluated. In some instances it will be sufficient to bound the problem based on source estimates derived from engineering data and then to determine whether the public exposure, evaluated from such data, was great enough to warrant further epidemiologic investigation. If further investigation is justified, then, in general, it will be necessary to develop the source term more precisely. This will require use of all relevant data to identify what was actually released, and when source term contributions from different streams must be quantified, their statistical ranges must be determined for each period of interest. The accuracy of the estimate must be determined as realistically as possible. No purpose is achieved by exaggerating the accuracy of the estimate; public confidence is better served by being candid about uncertainty. It is also important to use source term data in concert with information about potential environmental pathways. For example, if stored solid materials have not leached to groundwater pathways, further characterization of release to groundwater is unnecessary because no off-site release has occurred. Such a process sets up a feedback mechanism for increasing or decreasing the scope of the source term analysis. Finally, the source term, which is quantified from a combination of historic records and engineering estimates, can be further confirmed or supplemented by estimates derived from environmental pollutant concentration data and dispersion calculations.

Scoping studies are useful for determining the degree of effort required to conduct comprehensive source term studies, provided the goals of the dose reconstruction studies are clearly stated. A scoping study would use physical and engineering elements of the process or site and knowledge of operations over time to provide the initial input to a decision about whether to conduct an epidemiologic investigation. By its nature, a scoping study should be a conservative estimate that is used to preclude making an incorrect decision not to investigate the consequences of a particular release, either in terms of individual radiation exposures or as an epidemiologic study. Scoping studies also can provide guidance about which releases and pathways are of greatest importance in a more comprehensive analysis.

One approach to the initial phase of a comprehensive source term study is to survey all available records and to extract and evaluate all relevant data. This is a time-consuming and labor-intensive effort, requiring the locating of records that often date back many decades and are in most cases poorly indexed or identified. It also requires trained and technically competent personnel to review many documents to identify and catalogue any significant information. Because of the sheer number of plant records for some sites, a searching strategy must be developed for the retrieval of all relevant data. The credibility of a comprehensive source term study depends on confirming that all pertinent documents have been seen and evaluated. Complete records are essential in identifying the source term.

The major advantage of a comprehensive search is that it can eliminate bias in use of the data by allowing interpretation of all available data to determine the elements and conduct of the analysis of source term parameters, pathways, and persons exposed. A comprehensive search can be especially valuable when classified or proprietary information is involved. The major disadvantage of a comprehensive search is that it delays decisions on major elements of dose reconstruction and epidemiologic studies until after the data collection has been completed, and well after a significant expenditure of time and resources. Scoping studies conducted with care and public sensitivity could offset this potential disadvantage. Selective searches tend to be more efficient and less costly and for the most part provide the necessary data, because often much is known about the operations and history of the facility or site selected for study.

Regardless of the method used to estimate the source term, it is essential that the process involve public dialogue, especially to address public reports on any events observed, remembered, or suspected. Each event should be investigated and its contribution to the source term estimate included, either as a documented release or as a gap in the data set that is bracketed by other analyses.

Proprietary or classified information must be reviewed. When exposure reconstruction is indicated for a site or facility, it is essential that the reported source term and the data upon which it is based be removed from the arena of proprietary information, including security classification, if the study is to satisfy public scrutiny and achieve acceptance. This means that decision-makers will need to strike a balance between potentially conflicting goals: protecting proprietary or security information and achieving public confidence in the analysis performed. In most cases it will be possible to make essential information available and to explain why the information is sufficient for the estimate. One effective method, when some material cannot be widely disseminated, is to have a team of people with the appropriate security clearance review the classified or proprietary information and report to the advisory or steering group whether that information is or is not essential to the source term estimate. It also is possible to agree on a cutoff date beyond which proprietary data would not be essential because of the availability of alternative confirming environmental or release measurements.

DATA REQUIREMENTS FOR SOURCE TERM ANALYSIS

For a study to reconstruct a source term, all aspects that affected the releases must be considered and no relevant data can be overlooked. Because of the sheer quantity of documents that exist for some sites, the search process, whether selective or comprehensive, must ensure that all relevant documents pertinent to the period of interest are inspected and all useful information is extracted. The process should, to the extent feasible, obtain information from at least two different sources, such as measurements and process inventory, to ensure independent confirmation of source term data. Independent confirmation of release data should be given a high priority.

As far as possible, it is desirable to use original data, such as monitoring logs, rather than official summaries or reports. This might entail some judgment on the consistency or accuracy of records derived from different sources, but the reconstruction process itself should not attempt to edit the information unless there are obvious errors. The process of reconstructing data to fill in gaps should assess whether the contribution will be of epidemiologic significance.

To assist in the modeling of the environmental transport of the released material, it is often important to include data about the physical and chemical characteristics of the released material. Unless the process or plant has changed over time, this information can be included generically in the source terms.

Short periods of missing data can be bridged by interpolation between periods for which discharges are known. Longer periods might require extrapolation of release rates using changes in plant operations as a guide. For releases that were never measured or that were sampled only occasionally, it might be necessary to develop a model of the release that is related to the production process that caused it. The form of the model will likely depend on the data that are available to guide the calculations. Monte Carlo calculations can be used to derive best estimates, uncertainties, or probability distributions that characterize the particular source term.

Points of releases are particularly important; these should be established from plans of the site and facilities, information on processing activities, process flow sheets, and facility drawings. On-site inspection and mapping of the premises and remaining facilities is an essential part of such determinations.

Airborne releases of contaminants occur at a variety of locations, depending on the facility and the nature of its processes. Most routine releases of toxic chemicals or radionuclides have been discharged through chimneys or roof vents designed to exhaust process areas or equipment. The exhaust paths often used filtration systems to retain the pollutants of concern. However, other pathways, such as open doors, laboratory hoods, and air conditioners, could have contributed to emissions.

Atmospheric releases require specification of the source term from the height, diameter, air flow rate, and temperature in the stack and the contaminant concentrations in the discharged air. Information about the size distribution and chemical form of discharged aerosols and particles is important. Chemical forms of discharged gases also are important. Releases at ground level and in the open are characterized by release rates, by an effective height of the release point, and by information on chemical form and particle size, as appropriate.

Uncontrolled processes, such as burning of contaminated materials, also lead to atmospheric contamination. Other examples of uncontrolled processes are evaporation of volatile solvents spilled outside or discharged into a surface liquid waste stream and resuspension of contaminated particles from waste disposal pits or dried-out areas surrounding disposal ponds.

Liquid releases can occur from direct piping of waste products into a body of surface water or to a discharge point below the surface of a river, lake, or ocean. Disposal wells have been used to carry liquids into aquifers or to develop a water zone perched above an aquifer, and semisolids have been pumped into disposal ponds, lagoons and hydraulic fracture zones from which migration has occurred. In the latter case, adsorption of pollutants might have taken place in the soil column above the aquifer. Unlined disposal ponds and settling basins that were used to reduce the quantity of pollutants that reached aquifers or downstream surface water bodies, might have produced contamination of underlying soil layers.

Liquid effluent discharge rates and contaminant concentrations are two primary characteristics that should be documented. The fractions of the release that are dissolved and suspended are also important and should be identified if possible. Knowledge of the chemical forms of some pollutants can be critical to the exposure assessment. When liquid releases are initially directed to an on-site impoundment, assessing the off-site release requires documentation of the flow rate and concentrations leaving the impoundment. Similar considerations apply for evaluating concentrations of contaminants that enter an aquifer after they pass through a column of soil or rock above an aquifer.

Solid material releases have included a variety of contaminated materials disposed of in pits or trenches. The method used to emplace the waste and the cover provided influence the amount of airborne contamination generated during or after disposal, potentially leading to soil contamination. Spills of process chemicals or radioactive materials also have produced soil contamination, part of which could remain to the date of the study and could be measured for confirmation of the release pathway.

The physical and chemical forms of contaminants in solid wastes are important determinants of their environmental transport following emplacement. Some contaminants are incorporated in solids, in ion exchange resins, or in finely divided process waste materials. Knowledge of the distribution of the contaminants and the leachability of the waste form is highly desirable. Particle size is particularly important because of potential airborne transport.

EPISODIC RELEASES

It is important to distinguish episodic releases that justify special treatment in dose or exposure assessment, from the continuous releases at a facility. Episodic releases are those during which the release rate was at least 10 times greater than the average monthly or yearly release from the facility and that lasted fewer than 10 days. While use of average dispersion parameters is considered appropriate for high release rates of longer duration (more than 10 days), specific meteorologic or transport parameters for the time of the episodic events should be established and used if possible.

SOURCES OF INFORMATION

As previously stated, three independent data sets can be used to determine the source term: historic records, engineering estimates, and environmental monitoring data. Use of historic records involves collecting information on releases from stacks and other discharge structures or effluent pathways. This information can be partially confirmed or supplemented by engineering estimates, which are derived primarily from process information and environmental control technology performance. The resulting source term, based on a combination of historic records and engineering estimates, can be further confirmed or supplemented by estimates derived from environmental monitoring data. In some cases, the source term can be reconstructed by use of environmental pollutant concentration data and dispersion calculations.

An important first step in the source term estimation process is to locate measurements and estimates made by plant operators during the period of interest. After 1960, such estimates were frequently documented in environmental monitoring reports. More detailed information is available for years after the passage of the National Environmental Policy Act of 1969. Routine reports of release estimates are sometimes available in the files of a facility's health physics or industrial hygiene department. However, it is recommended that the formal reports of effluent releases be validated by examination of the original, handwritten logbooks, files, or analytic data sheets that contain sample mass, activity, or concentration measurements. These original data also can indicate, directly or indirectly, the uncertainty associated with the measurements. Daily waste-processing logs often contain detailed information about transfers of liquid effluents.

Unplanned releases may be documented in several ways. The least formal source is found in routine production reports or in periodic reports of the cognizant safety organization. Events involving larger releases are usually discussed in some detail in incident reports or accident investigation reports. In recent years, data bases of unusual occurrences have been established and updated routinely. Very large releases and their causes are almost certain to have been evaluated by an investigation team and formally documented.

Interviews with current and former employees who witnessed or investigated such events can provide useful additional information. These persons often have information not found in formal reports. They also can offer information about routine plant operations and about incidents of lesser consequence that might have occurred more frequently.

For some chemical releases, it is possible to estimate the source term by using logbooks of material usage or transfer and information about the physical and chemical characteristics of the process and effluent treatment systems.

For long-lived radionuclides or chemicals, source terms can be "back calculated" using transport models from measurements of soil or sediment deposition, if such data are available. The magnitudes of episodic releases can be estimated from measurements of the contaminant in air, water, or biota during or shortly after the event. The success of these approaches depends on the quality of the data and the transport model used in the calculation and on the mobility and persistence either of the contaminant of interest or of a surrogate tracer in the environment.

Some of the information obtained in source term studies is useful for model development, especially data found in production files, specific process data, purchasing records, materials accountability reports, and logbooks of material use or transfer. In some cases, detailed information on reactor operating cycles and fuel cooling times will be needed to quantify estimates of short-lived nuclides. Other models will not require such detailed information. In many cases, operational data will no longer be available and the modeling effort will reflect the limitations imposed by the data base.

Research and development reports that describe prototype facilities and their operation can provide useful information for understanding releases from the process equipment. These reports typically describe initial production runs and the problems that were encountered. Measurements of release fractions during pilot plant or early start-up operations are particularly useful, but records of release fractions of particles and volatile radionuclides measured in later years can be helpful as well. Reports that describe effluent treatment systems or changes in them are especially valuable, as are operational measurements.

BIAS AND UNCERTAINTY IN RELEASE ESTIMATES

Release estimates inevitably are subject to uncertainty and can be biased because of unrepresentative effluent sampling or because of the occurrence of accidents in which material was released under less-than-ideal conditions for measurement. For example, a collected sample might not have been representative if the effluent discharge stream or the contents of the holding tank being sampled were not well mixed. For airborne effluents, another source of possible bias is unequal rates of sample withdrawal. Biases can be positive or negative, and they can lead to over- or underestimation of the releases. Corrections for such biases can be inadequate if information about the relevant parameters is limited.

Lack of monitoring data is another source of uncertainty in release estimates. In some cases, releases were not measured, and sometimes the records of measurements have been lost or destroyed. Data might be missing for only a brief period because of a sampler failure or there can be substantial uncertainty in the release estimates that must therefore be developed without effluent-monitoring data. Incomplete information about processes and about the historic details of facility operations also contributes to uncertainty.

Even when the samples extracted were representative of the effluent streams, there are other sources of bias and uncertainty. Part of the sample might have been lost in the sampling line, leading to an underestimate of the release. Incomplete collection of the sample by the filter also would cause an underestimate.

Uncertainties also arise when estimating correction factors for line loss and collection efficiency. Uncertainty also can come from the measurement of the quantity of material collected in the sample and of the volume of the sample. In many cases, only the analytic measurement uncertainty was recorded and included in reports of releases. This sometimes changed over time because of improvements in measurement equipment and in the development of new analytic techniques that led to more isotope-specific analyses.

GAPS IN RELEASE DATA

For some periods of time or for some types of operations, production, release, or emission data might not be available. This could result from loss of documents, destruction of documents considered obsolete, or from security concerns. Gaps in information should be filled, as far as possible, by extrapolation of release data obtained for comparable operating periods or by reconstruction from other data such as exposure records, environmental monitoring results, or waste shipment records. In some cases steady-state conditions of operations and releases will be sufficiently well established to permit interpolation or extrapolation to undocumented periods. Whenever this is done, the methodology used should be fully described and justified.

Fundamental physical principles can be used when the facility's processes are well understood. If a reactor was the original source of the release, fuel burn-up calculations can be made to estimate core inventories, and releases can be estimated once the cooling time is determined and the separation process is described. This methodology was used successfully for reconstructing ^{131}I releases from the Hanford, Washington, reactor (Heeb and Morgan 1991). For simpler situations, release fractions can be estimated from records that detail the throughput of the material, its physical and chemical characteristics, and the facility's effluent treatment systems. Research and development reports can provide facility descriptions, information about effluent control technology, and plant operation data.

SUMMARY AND RECOMMENDATIONS

The source-term estimate reported in a dose reconstruction study should provide information about the quantities of materials released in a form that is suitable for the environmental pathway models used. For this reason there should be a connection between the source term analysis and the development of the pathway models so that the arrangement and presentation of the various source streams will be appropriate.

Because the data analysis might lead to a decision not to pursue further study of some minor releases, it could be important to report their magnitude and when they occurred. The same could apply to releases that mostly affected the plant site itself. In a detailed study, completeness of data and comprehensive estimates of reported releases should be the goal of the reported source term; this is ensured by using all relevant data, by investigating all reported events and data gaps, and by determining each release value by as many independent means as possible. The following six recommendations are made by this committee to ensure that the source term is as complete as necessary for the goals of a study. The source term often provides the foundation of a dose reconstruction study.

8. Scoping studies should be the primary approach to initiating a source term evaluation. These should be followed, if appropriate, by more comprehensive studies. The scoping study of the source term should seek to generate the data needed to identify the environmental pathways of potential importance and to permit estimation of the concentrations of radionuclides to which the public might have been exposed.

9. To ensure maximum confidence in the source term analysis, proprietary or classified information should be made available for the analysis, or a mechanism should be developed to determine whether such data are essential to the accuracy and consistency of the analysis.

10. The source term should be derived chiefly from available original data in as many different ways as practical. The source term should be confirmed, wherever possible, by comparison with independent environmental monitoring data from another source of information.

11. To ensure completeness and accuracy of the estimated source term, all relevant data should be evaluated. Any gaps in the data should be analyzed carefully for their significance and filled by reconstruction from existing data if appropriate.

12. Episodic events should be documented as separate releases for

specific consideration in environmental transport and dose calculations. An event is considered episodic if it lasted for less than 10 days and if the release rate was at least 10 times the average monthly or annual rate.

13. The release quantities provided for use in a comprehensive study of the source term should be complete, unbiased estimates of all amounts and forms of relevant materials released to the environment.

4

Environmental Pathways

HUMAN EXPOSURE TO radionuclides occurs as a result of transport along various environmental pathways. It is the overlap in space and time of the "region of influence" of decaying radionuclides with the presence of a person that results in radiation exposure of tissues and organs. The objective of an environmental pathway analysis is to estimate human exposure rates or to determine radionuclide concentrations in air, water, and foodstuffs. This information is used in the assessment of radiation doses to representative or specific individuals.

The environmental pathway analysis, which in most cases uses the source term assessment as an input, may involve an iterative procedure for making estimates of exposure rates and of environmental concentrations of radionuclides. In the scoping study, which is carried out to determine if there is a need for a full-scale dose reconstruction study, only the most important environmental pathways and radionuclides are considered, and readily available information on the characteristics of the site and on the population distribution is used to estimate the exposure rates and environmental concentrations at a limited number of locations of interest. In a full-scale dose reconstruction study, the number of estimates of exposure rates and environmental concentrations is gradually expanded, and their quality is improved as: (1) the full source term is taken into consideration, (2) the results of detailed studies of the characteristics of the site with respect to the environmental behavior of the released radionuclides are used, (3) all environmental pathways that

could lead to radiation exposures are considered, (4) all available historical measurements of environmental radiation are used to bypass environmental transfer models or to validate those models, and (5) the uncertainties attached to the results of the environmental pathway analysis are
estimated. It is important to have epidemiologists involved in the full-
scale dose reconstruction study in order to ensure that the information
developed is appropriate for epidemiologic decisions and planning. The
purposes of this chapter are to describe the environmental pathways that
need to be considered in dose reconstruction studies and to list criteria for
their assessment.

TRANSPORT OF RADIONUCLIDES AND OTHER CONTAMINANTS

Humans come into contact with radioactive and other materials by
means of a variety of pathways, and the movement of the materials can be
affected by many physical, chemical, and biologic processes. An important feature of any dose reconstruction must be a critical analysis of pathways so that those of most relevance can be identified. The environmental releases most frequently encountered in dose reconstruction studies
are direct releases to the atmosphere or to the hydrosphere; they are
considered in turn.

Direct Releases to the Atmosphere

Analysis of the environmental pathways from atmospheric releases
to humans requires study in four areas. First, the meteorologic processes
that govern atmospheric dispersion and the precipitation processes that
deposit gaseous and particulate emissions must be considered. Second,
the pollutant concentrations in ground-level air must be quantified. Radiation exposure by inhalation and exposure to external irradiation from
the radioactive materials present in the cloud are derived at this stage.
Next, the amounts and the physical and chemical forms of the materials
deposited on the ground need to be determined. This information can be
used to estimate exposures to external irradiation and, occasionally,
through ingestion of soil. Finally, the behavior of radionuclides after they
are deposited on the ground must be ascertained. They can be resuspended in the air; dissolved in groundwater; or enter the food chain
through agricultural processes, through irrigation systems into crops or
into the drinking-water supply.

Radioactive materials from sources such as nuclear power plants are
typically released between the ground surface and an elevation of 100 m
into a region of the atmosphere called the planetary boundary layer. The

elements of the airborne plume are affected by turbulent eddies in the layer that diffuses the effluent material as the plume is transported downwind. Generally, the combined influences of diffusion and transport are called dispersion (Brenk and others 1983). Extensive discussions of atmospheric transport processes can be found in *Meteorology and Atomic Energy* (Slade 1968) and the *Handbook on Atmospheric Diffusion* (Hanna and others 1982).

If the emission rate is known and approximately constant or randomly distributed in time (or if it can be inferred) and if the average or specific meteorologic conditions are known, then the movement of the materials can be described by a variety of models of atmospheric transport (Brenk and others 1983). The object of the atmospheric transport models is a description of the ground-level concentration of materials as a function of time. Unfortunately, atmospheric modeling is subject to a great deal of uncertainty, and it is preferable, therefore, to have actual measurements of the airborne concentrations wherever possible.

The presence of radionuclides in air may lead to two modes of radiation exposure—inhalation and external irradiation. If the individual considered is outdoors, the measured or calculated radionuclide concentration in outdoor air at that location is used without modification in the estimation of the radiation dose, using an appropriate model for the inhalation dose. However, people spend most of their time indoors, and indoor concentrations of most air contaminants are as a rule smaller than are the outdoor concentrations because of the filtration provided by buildings. It is frequently assumed, on the basis of limited experience, that indoor air concentrations of radionuclides in particulate form are about one-third those found outdoors if building leakage is the only migration path.

If the individual is outdoors, the measured or calculated radionuclide concentration in outdoor air at that location can be used without modification in the derivation of the external radiation dose. If the individual is indoors, the structure of the building attenuates the radiation emitted by radionuclides in the outdoor environment. The magnitude of the shielding factor is highly variable (typically between 0.001 and 0.5) because it depends, among other factors, on the type of building and on the floor level considered. The exposure geometry varies according to whether the irradiation is due to a contaminated overhead plume or whether the individual is immersed in a near-ground contaminated atmosphere.

Also, people do not spend all of their time at the same location. Information on the whereabouts of people is particularly important in the aftermath of an accident. If countermeasures, such as relocation, were taken, knowledge of the radiation field distribution along the routes of relocation and at the relocation site is needed. These are important factors in the determination of the total exposure of the relocated populations.

Particles and volatile matter released as airborne effluents can be deposited on vegetation and the ground surface. Wet deposition processes include rainout and washout. Rainout occurs when the pollutant becomes involved in precipitation formation processes within a cloud and is subsequently removed from the atmosphere with the precipitation. Washout is the removal of the effluent from the plume by entrainment in falling precipitation. In the absence of precipitation, effluent material also can be removed from the atmosphere through gravitational settling onto the ground, vegetation, or other ground cover, such as buildings.

The transfer of airborne contaminants from ground-level air to the ground surface, including vegetation, is usually modeled through the use of the deposition velocity concept, which is the quotient of the deposition of radioactivity (in becquerel, Bq) per unit area (in square meters, m^2) to the ground surface and of the time-integrated concentration in ground-level air (Bq/m^3).

Choosing an appropriate value for the deposition velocity is difficult and can be a major source of uncertainty, possibly of four or five orders of magnitude. However, use of consistent methods of soil sampling and analysis must be documented for such data to be considered reliable. Deposition velocity varies as a function of particle size, wind speed, surface conditions, and most significantly as a function of the physical-chemical form of the contaminant (Sehmel 1980). The presence or absence of rainfall also can be crucial, and separate methods of predicting the washout or rainout of contaminants as a function of physical-chemical form must be used (Engelmann 1970). Wet deposition also can occur as a result of "washout" of these components by rain and other forms of precipitation. Dry and wet deposition are "integrating pathways"; the areal amount of deposited material is proportional to the time integral of the airborne concentration. Such processes can lead to localized areas of higher ground deposition or "hot spots" resulting in higher dose rates and doses. It is difficult to determine where and when hot spots occur, particularly in retrospect, because it requires knowledge of releases and concomitant atmospheric conditions (stability, wind speed, and the geographic patterns of precipitation such as rain, sleet, hail, fog, or snow).

Both wet and dry deposition depend on the physical-chemical form of the radionuclide. The aerodynamic diameter of particles significantly affects deposition rates; higher deposition rates occur for large particles >0.2 µm (gravitational settlement) and for very small particles <<0.2 µm (Brownian motion). Particle size can change with time (and distance) because of sorption on atmospheric particles, preferential depletion from deposition or washout, or agglomeration (Megaw 1965). Reactive gases such as elemental iodine (I_2), can be preferentially deposited on vegetation and ingested by grazing animals. Particulate iodine deposits and organic iodides, such as methyl iodide (CH_3I), have even lower deposition rates. Soluble gases and larger airborne particles are more subject to washout by precipitation than are inert gases or small particles.

If a single contaminating event has taken place and if measurements have been made (such as external gamma exposure rate or deposition of one or more radionuclides or stable materials), it is often possible to begin the dose reconstruction with such data and to bypass the need to model the transport of radionuclides up through this stage. Even if the contamination has been chronic, it is often useful to sample the soil to measure the longer lived components of the contaminant materials and to infer the deposition of the shorter lived components. The alternative is to depend on atmospheric transport and deposition models, which are much less reliable than are direct measurements. The use of soil-monitoring data is particularly valuable when the longer lived contaminant has the same chemical and physical properties as do the shorter lived components.

The presence of radionuclides on the ground can lead to human exposure by external irradiation. Here again, if the individual considered is indoors, the structure of the building acts as a shield against the radiation emitted by radionuclides present in the outdoor environment though there can be cumulative adsorption or deposition on the roof.

Radionuclides deposited on the ground can become resuspended in the atmosphere through aeolian processes, that is, through the action of the wind. They also can be transported on the ground surface by runoff, or they can migrate vertically into deeper layers of soil.

Resuspension of deposited radionuclides by wind or mechanical disturbance (for example, by transport on moving vehicles or people) can lead to inhalation intake or increased deposition at downwind locations. The amount of resuspended material depends on the characteristics both of the radionuclide particles and of the ground surface (surface roughness and vegetative cover) as well as on atmospheric factors such as wind speed. Resuspension is generally a minor exposure pathway if releases from the primary source are continuous, but it can become relatively significant after the primary releases cease. Much of the research on resuspension mechanisms has been done in arid environments, but after the Chernobyl reactor accident in 1986 the extent of such processes also has been studied carefully in more humid ecosystems. Resuspension could be a major factor in the transport of contaminants from piles of mill tailings and from desiccated storage lagoons.

Most of the plutonium transported off-site from the Rocky Flats Plant in Colorado was in windborne soil that had been contaminated by leaking oil drums. This pathway was unsuspected by the operators of the plant

(Hammond 1971), and demonstrates the need for vigilance in considering possible modes of transport. Although there are models that might help researchers to predict the effects of windborne transport (Langer 1989), the process is so complicated that it is preferable to use measurements to infer the source term or to describe the starting point for use in the dose reconstruction (Hardy and Krey 1971).

On-site contamination of soil and other surfaces has been found at many locations. During periods of normal rainfall, some fraction of the contaminants can be carried off-site by surface runoff. For example, this is a factor that may have influenced the estimates of uranium depositions between two studies at the Feed Materials Production Center in Fernald, Ohio (Voillequé and others 1991, Stevenson and Hardy 1993). In other cases, such as at Mound Laboratory, an episodic rain event was responsible for significant transport of radionuclides across the site boundary (Rogers 1975). Also, at Goiania, Brazil, runoff led to significant contamination of water bodies and sediments (Godoy and others 1991).

Especially for non-radioactive contaminants, such as trichloroethylene, and for tritiated water, transport by groundwater is often a major consideration. In many cases, the movement of nitrates and phosphates is significant. Materials that have been deposited or spilled on the surface soil are transported through the vadose zone (the unsaturated soil layer— the region above the permanent groundwater aquifer) and eventually through groundwater. This is a pervasive problem at Department of Energy sites. The possible interception of contaminated plumes by wells that supply drinking water is one of the more challenging pathways to be modeled. Because many of the factors that control transport are poorly known, it is preferable to depend on actual concentration measurements of contaminants instead of on modeling.

Radionuclides can enter the terrestrial food chain by direct deposition onto forage and food crops or by plant uptake from soil or from irrigation water. Contaminated plants can then be consumed directly by humans or ingested by animals with transfer of the radioactivity to human food products (meat, milk, or eggs). Animals can also inhale airborne contamination. Fortunately, plutonium and other transuranic actinides are relatively insoluble in water and their uptake by plants and animals via water pathways is very low.

Estimating deposition onto plants is similar to estimating total deposition, except that a retention factor is included to estimate the fraction of the total deposition that adheres to vegetation.

Investigation of plant uptake of radionuclides from soil requires considering the food crop, the edible portion (underground roots and tubers tend to show less radionuclide uptake than fruits, for example), soil characteristics, and farming practices (liberal use of fertilizers can reduce radionuclide uptake by crops).

Human exposure occurs through the consumption of contaminated foodstuffs from the natural environment (such as berries, mushrooms, or wild game) or in agricultural foodstuffs (such as green vegetables, milk, or beef). Radionuclide concentrations in the consumed foodstuffs can differ substantially from those in the raw agricultural products at the point of production. For example, leafy vegetables contaminated by direct deposition of radionuclides exhibit greater concentrations on the outer leaves, and removing those leaves as well as washing the vegetables will eliminate a substantial fraction of the contamination. Boiling or frying food also can reduce radionuclide concentrations.

People do not necessarily consume local agricultural products. Even common foodstuffs like milk or vegetables often are transported large distances from the site of production to the point of consumption. Although the movement of foodstuffs must be accounted for, it is important primarily if it results in a reduction in the consumption of contaminated food by the local population, particularly in the case of milk. There are also seasonal variations in uptake via the food chain. Grazing practices are site specific and can include the seasonal movement of grazing animals from one pasture to another.

Finally, countermeasures after accidents can lead to drastic decreases when highly contaminated foodstuffs are removed from commerce and replaced with clean products. Even if contaminated foodstuffs are not removed, people often voluntarily avoid these foodstuffs and change their dietary habits to reduce their intake.

Direct Releases to the Hydrosphere

Waterborne process effluents are often released to water in rivers, lakes, or seas. Another source of emission to waterborne pathways is through sanitary sewers. Although this is not likely to be a source of exposure to the public, the potential should not be overlooked, in particular because sewage sludge is sometimes converted to agricultural fertilizer. Sanitary waste is usually released to streams after biologic digestion and chlorination, leaving the dissolved radionuclides in solution.

The steps involved in the analysis of the environmental pathways from releases into the hydrosphere to humans are conceptually similar to those related to atmospheric releases. They entail the determination of the pollutant concentrations in locations where water is drawn for drinking or irrigation, the estimation of the amounts deposited on sediments, which can lead to exposure by external irradiation, and analysis of the subsequent behavior of radionuclides transported by water or deposited on sediments.

In general, discharges to fresh water pose a potentially greater threat to public health than discharges to the marine environment because there is less potential for dilution in freshwater; freshwater is used for drinking and irrigation; and for most radionuclides, bioconcentration factors are higher in fresh water than they are in marine waters.

Aquatic dispersion depends a great deal on the nature of the body of water (pond, lake, river, or saltwater body) that receives the effluent. In particular, radionuclides will accumulate in closed water bodies, such as small ponds and lakes. The nature of the receiving water body is also significant in determining the amount of contact aquatic organisms have with the radionuclides.

The concentration of radionuclides in freshwater systems can be modeled as a function of time (Jirka and others 1983) in a fashion similar to that for atmospheric releases. The major pitfall in models of freshwater systems is uncertainty in the amount of dilution between the release point and the downstream location of water withdrawal. The possible occurrence of stratification, or the lack of mixing of sections of water, can cause uncertainty as well. These uncertainties are particularly pronounced for episodic releases, because it is not always possible to determine whether the plume bypassed or was captured at the water withdrawal location. Again, the use of actual measurements of concentration is preferable to the prediction of radionuclide concentrations by aquatic dispersion models.

An important factor to consider in the analysis of the environmental behavior of radionuclides is the partitioning between water and sediments in a stream bed. The behavior of radionuclides in freshwater, estuarine, and marine ecosystems depends on their physical-chemical form. Larger particles of insoluble materials can settle out near the discharge point. Radionuclides that were dissolved initially can precipitate as a result of chemical reactions in the receiving water body or they can be adsorbed on sediments or suspended particles (e.g., cesium on clay particles).

The chemical form of the radionuclide in an aquatic ecosystem can change with time and distance from the release point. Radionuclides can become less soluble because of chemical changes and precipitate onto bottom sediments. Physical changes in the receiving waterway may also affect radionuclide sedimentation. Widening of a river can result in lower flow velocities and increased sedimentation, as will the presence of impoundments such as dams. Changes in the chemical composition of the waterway, such as occur in an estuary, can affect radionuclide sedimentation. In particular, increased radionuclide deposition has been observed in the vicinity of the freshwater-saltwater interface as a result of a change of ionic potential, and adsorption on benthic organisms can be significant in brackish or estuarine waters.

Better characterization of the environment and of the physical-chemical form of the effluents will allow more accurate estimates of the radionuclide concentrations in aquatic foodstuffs and reduce uncertainty.

Radionuclides in water and sediments can be absorbed by aquatic organisms that are part of the human food chain, although the extent of absorption varies with the form of the radionuclide because changes in form affect the transport and uptake of the radionuclide. For example, radionuclides in particulate form are especially subject to concentration by shellfish or other filter-feeding organisms. One might also have to consider cross-contamination of terrestrial farm animals through the use of contaminated fish bone as feed. The presence of chemically similar stable elements can greatly alter uptake of radionuclides by aquatic organisms and, consequently, their concentrations in the food chain. For example, the degree of uptake of ^{137}Cs by freshwater fish is inversely proportional to the potassium concentration in water, and the uptake of ^{90}Sr is inversely proportional to the concentration of stable calcium in water. Consequently, the bioaccumulation of cesium and strontium isotopes is much lower in marine systems than it is in freshwater because of the larger amounts of potassium and calcium in seawater.

Estimation of radionuclide intake from aquatic food chains requires an estimate of the radionuclide losses during food preparation. It is important that the characteristics of specific foods be considered so that all appropriate pathways are evaluated. (Small fish, such as smelts, are often consumed with their bones; larger fish are usually boned. The intake of radionuclides such as ^{90}Sr, which tends to concentrate in bones, is apt to be higher for consumers of small fish.)

The potential pitfalls in reconstructing doses associated with the discharge of radionuclides to the aquatic environment depend on whether the discharges were to the freshwater or marine environment and whether they were chronic or episodic. The factors that introduce large uncertainties into an aquatic dose reconstruction include retention or delay of groundwater migration in unsaturated subsurface flow, the removal of radionuclides by sedimentation processes and their resuspension during floods, and the bioavailability of radionuclides in solution and in the bottom sediment.

In all cases, the chemical form of the radionuclide and the chemistry of the receiving water and sediment greatly influence bioavailability. Also, the influence of high water levels after storms in remobilizing sediments and the possibility of contamination of agricultural lands by flooding might need consideration.

FURTHER CONSIDERATIONS

Even if there is an abundant base of monitoring data for environmental media, the estimates of exposure through some or possibly all of the environmental pathways usually require the use of mathematical models, i.e., quantitative approximations of the processes that affect the transfer of radioactive substances in the environment to the point at which there is human contact. The models are combinations of equations and parameters usually formulated in computer codes. The parameters require fairly precise numerical estimates if the model is to be more than heuristic.

Appropriate Use of Mathematical Models

When actual measurement data are either absent or incomplete, mathematical models are needed to estimate concentrations of radionuclides in air, water, soil, food, fodder, and human organs. Models are used to extrapolate information from situations where measurements have been made to situations where measurements are lacking. The model's complexity varies depending on the nature of the assessment question and the degree of resolution required to answer the question. Because mathematical models are merely representations of reality, any simplifications and assumptions inherent in their construction, implementation, and execution must be examined carefully through uncertainty and sensitivity analysis of the results. It is essential that the computer codes be properly verified to ensure the absence of coding errors. Every attempt should be made to validate the predictions of these codes against relevant but independent data sets. Caution must be exercised in the uncritical use of "off-the-shelf" assessment codes that have been developed for the purpose of regulatory analysis. Their equations and data bases support generic assessments for reference situations, but because they usually are designed to determine compliance with regulations, they are seldom applicable to realistic estimates of exposure. The user is forced to use the code as a black box and typically can change only the assumptions about parameter values. Because the user is denied access to the source code, structural modifications cannot be made to adapt the model structure to the unique situation presented at a given site (Hoffman 1993). Finally, although computer codes can be verified, peer reviewed, and sanctioned by specific government agencies, their results still rely on the professional judgment of the user, and different users might get different results.

Uncertainty Analysis

Pathway modeling of exposure should include quantitative estimates of uncertainty to measure the degree of confidence that can be placed in the exposure estimate. The dominant contributors to uncertainty should be identified, and this information should be used to acquire additional data to reduce uncertainty. Quantitative uncertainty analysis, however, requires a rigorous definition of the target end point of the assessment because different results will be obtained depending on whether the objective is to estimate exposures for representative individuals or for actual persons. If the latter, further information about age, gender, diet, lifestyle, and residential history is needed in the modeling process.

Probability density functions (pdf's) are typically needed to express uncertainty in model parameters for which the true value is unknown. In the absence of site-specific data, the derivation of parameter pdf's requires quantification of the subjective degrees of belief of the investigators or outside experts. There may be divergence among experts as to the best parameter pdf. The resolution of this issue will require the acquisition of new experimental data or the use of formal approaches for eliciting subjective information from groups of experts.

There are other issues of concern:

• It is necessary to distinguish between uncertainty that arises from unexplained variability in the observed data and uncertainty that results from a lack of knowledge about a true but unknown value.

• It is important to evaluate known dependencies among model parameters and their effect on the uncertainties in assessing dose and on the epidemiologic association between dose and health outcomes.

• It is also important to include the model's uncertainty itself along with uncertainty in model parameters. This might require resolution through sensitivity analysis.

Once uncertainty has been propagated through the exposure pathway equations, different methods can be used to express confidence in the model result. Some investigators have used a geometric mean and geometric standard deviation; others have used an 80% to 95% probability interval based on expert judgment; still others use the entire joint pdf of the model result. The one most appropriate for dose reconstruction depends on the type of decision that is to be made with the exposure information. While there is no standard for all situations, it is recommended that uncertainties be quantified as a confidence interval (IAEA 1989). If the uncertainty is due to natural variability, a pdf may be appropriate but confidence intervals about the pdf should be provided. Confidence statements should include all sources of uncertainty.

Several areas of uncertainty have been identified in past and current pathway analyses:

• It is difficult to estimate the precise wind trajectory fields for sites of episodic releases or those for which meteorologic data are inadequate.

• Determining the physical-chemical forms of material released and transported in air, can be problematic but is important because these forms can affect the rates of removal from the plume and the retention of materials in the human lung.

• It is difficult to estimate the amount of wet versus dry deposition and whether the deposited material is intercepted and retained on natural or synthetic surfaces, such as trees or buildings.

• Site-specific values need to be determined for the food chain transfer of radionuclides. Element-specific transfer coefficients and rate constants must be quantified, and estimates of the transport of contaminated foodstuffs outside or into the region of concern must be made.

• Accounting for unusual pathways of exposure, such as the contamination of cisterns through wet deposition, and the uptake of possibly high levels of contamination by wild game and waterfowl can be difficult.

• Correctly quantifying the attributes of the human receptor is another area of concern. The locations of residence and work, recreational activities, dietary habits, characteristics of metabolism, state of health, and the effectiveness of shelter all must be considered. Many additional attributes must be properly quantified to translate the estimate of exposure into an estimate of dose and an estimate of health risk. These issues are more significant in the process of obtaining exposure estimates for actual persons than for representative (hypothetical) individuals.

SUMMARY AND RECOMMENDATIONS

One important feature of any dose reconstruction must be a critical analysis of all possible environmental pathways to identify those of most relevance to the population and to those special groups that might have been the most exposed. The environmental releases most frequently encountered are those directly to the atmosphere or to the hydrosphere. Conceptually, the steps involved in the analysis of the environmental pathways from releases into the atmosphere or the hydrosphere to humans are similar. They entail the determination of the pollutant concentrations in locations where the radionuclides could be inhaled or ingested; the estimation of the amounts deposited on the ground or in sediments, which can lead to exposures by means of external irradiation; and the analysis of the subsequent transport of the radionuclides through physical or biologic processes that will bring the contaminants into contact with humans.

The committee makes four recommendations:

14. **Insofar as possible, measurements of environmental radiation or of radionuclides should be used in the environmental pathway analysis.** For example, if a single contaminating event has taken place and if measurements have been made (such as external gamma exposure rate or deposition of one or more radionuclides or stable materials), it is often possible to begin the dose reconstruction without the need to model the transport of radionuclides up through this stage. Even if the contamination is chronic, it is often preferable to take suitable soil samples to measure the longer lived components of the contaminant materials (such as ^{137}Cs or ^{129}I) and to infer the deposition of the shorter lived components (such as ^{131}I), rather than to depend on atmospheric transport and deposition models, which are much less reliable than are direct measurements.

15. **Even if there is an abundant base of monitoring data, mathematical models are usually needed to extrapolate information from situations where measurements have been made to situations where measurements are lacking.** Every attempt should be made to validate the predictions of the models against relevant data sets. Caution should be exercised in the use of "off-the-shelf" computer codes that may have been developed for other purposes such as regulatory analyses.

16. **Environmental pathway analyses should include quantitative estimates of uncertainty to indicate the degree of confidence that can be placed in exposure estimates.**

17. **For accidents, there should be careful scrutiny of any countermeasures, such as removal of contaminated foodstuffs from commerce.** Even if contaminated foodstuffs are not removed, people often voluntarily avoid contaminated foodstuffs and change their dietary habits. For routine releases, attention should be paid in the assessment of ingested radionuclides to the movement of foodstuffs into the region of interest because people do not necessarily consume local agricultural products.

5

Radiation Dose Assessment

INFORMATION ON THE PATHWAYS through which radionuclides are transported to the public from a contaminated site provides the starting point for calculating the dose to individuals and for estimating health risk. This chapter assumes that ambient exposures from an overhead plume or from radionuclides deposited on the ground have been determined and that the concentrations of radionuclides in air and water in the vicinity of the exposed population have been estimated. It is further assumed that both have been described in quantitative terms suitable to the needs of dose estimation. This chapter describes the process of dose estimation, the different types of dose assessment that can be undertaken, and the uncertainties involved.

SOURCES OF EXPOSURE

Three sources of radiation exposure that must be taken into account in estimating the dose either to representative or to specific persons are: (a) external exposure due to submersion in contaminated air or due to radiation from an overhead plume or from radionuclides deposited on the ground, (b) the inhalation of radionuclide-contaminated air, or (c) the ingestion of radionuclides in water and foodstuffs.

Ambient Exposure

External exposure to ambient radiation may be a small contributor to the total dose of a member of the public in many dose reconstructions. However, external exposures can be significant for noble gases such as ^{41}Ar or ^{133}Xe or long-lived radionuclides that are deposited on the ground. This exposure may be the easiest to estimate, albeit with some uncertainty, if suitable environmental measurements were made in the open and in typical shelters. Absorbed doses from external sources have been calculated for an extensive list of radionuclides and tabulated in Federal Guidance Report 12 (Eckerman and Ryman 1993). These coefficients are based on detailed scattering calculations from the distributed source to the organs and tissues in a mathematical phantom. Since the coefficients are specific for each radionuclide, it is essential to consider the ingrowth of decay products. For example, an assessment of the external exposure from ^{137}Cs must consider the ingrowth of ^{137m}Ba.

Inhalation Exposure

Contamination of the air with radionuclides can be described in several ways, including the concentration of radionuclides in the inhaled air, the size of the inhaled particles, and the solubility of the particles that contain the nuclides. Calculation of the dose to the whole body or to a specific organ requires further information: the estimated duration of daily exposure (in hours), the breathing characteristics of that individual, a model of the respiratory tract that allows calculation of the amount of airborne particles deposited in the airways, the physical and metabolic characteristics of the deposited radionuclide, and the size of the individual. Aerosol particles are usually assumed to be lognormally distributed in terms of their aerodynamic diameter and hence can be characterized by their geometric mean and standard deviation.

The models currently used for determining the deposition of airborne contaminants in the respiratory tract are being revised (ICRP 1995). The new versions will be more detailed and will define the fate of a deposited radionuclide through modeling the clearance characteristics of the particular inhaled compound. If a particle is insoluble in body fluids, it is assumed to clear slowly from the location in the respiratory tract (primarily the lung) where it is deposited. If the particle is soluble, it will be cleared from the lung more quickly. Once cleared to the blood or the gastrointestinal tract, the radionuclides can translocate to other organs before finally being excreted from the body. This process can take several months or longer. The chemistry of the compound containing the radionuclide strongly influences the biodistribution of the radionuclide and the transfer rates. Alkaline earth elements such as strontium or radium that are similar in ionic radius and in chemical characteristics to calcium will deposit in the skeleton. Some of the actinide elements (those elements with an atomic number equal to or exceeding actinium's 89) also will deposit in the skeleton, primarily through their interaction with organic molecules before they are mineralized in bone. An example of this is plutonium. The actinides also tend to accumulate in the liver. As a result of this accumulation, measurements of radionuclides in tissue samples may provide a method of dose assessment when long-lived radionuclides are involved.

The absorbed dose due to inhaled radionuclides can be calculated by means of an appropriate model that takes into account lung deposition, transpiration, organ distributions, and clearance rates (Loevinger and others 1991). The absorbed dose to an organ is defined as the radiation energy absorbed per unit mass of that organ; it is determined from the fraction of energy absorbed by the organ from radiation emitted by radionuclides (alpha particle, beta particle, gamma or x-ray, etc.) in particular organs or in adjacent organs. The deposition of 100 ergs/gram is one rad; 100 rad is a gray (Gy). The unit of dose equivalent is the rem, which is the absorbed dose in rad multiplied by a modifying factor, or quality factor, to adjust for the biological effectiveness of a particular type of radiation. The SI unit of radiation dose equivalent, the sievert (Sv) equals 100 rem. Many body organs in addition to the lung may be exposed to the radiation emitted by the inhaled radionuclide. Both doses to specific organs and equivalent doses to the body can be estimated.

In the decades before 1970, models of the relationships between physiologic processes and dose were simplistic. Nonetheless, they led to radiation protection strategies that were sufficient to limit the potential health hazard to exposed persons. This simpler perspective has evolved more recently into approaches (e.g., ICRP 1989, 1991, NCRP 1993) that translate organ doses into an equivalent dose to the whole body (an effective dose) by weighting the contributions of all organ doses. Committed doses, i.e., the total dose delivered over an extended period of time, typically 50 years following intake, are used for this purpose. However, for the purpose of epidemiologic study, annual organ doses for low-LET and high-LET radiation are preferred to effective doses. For exposures to long-lived radionuclides, the calculation must be a time-dependent one.

Ingestion Exposure

The procedure for calculating radiation dose from ingested radionuclides is similar to that for calculating the dose from inhalation. Ingestion is defined as the swallowing of radionuclides in water (or other liquids), and in food. Absorption of radionuclides from the gastrointestinal tract to blood is affected by such factors as chemical form and interactions with other food in the gastrointestinal tract. Given such variable absorption, the most practical means of estimating the dose is to use a model with average absorption kinetics and to calculate the dose to organs based on normal physiological processes. Inhaled particles brought out of the lung by ciliary action are also ingested. From that point, the radiation dose is determined as if the radionuclide were ingested, not inhaled. The radiation dose to the gastrointestinal tract is largely due to the radionuclide activity within the contents of the bowel, the critical organ for some major fission products, for example, ^{103}Ru and ^{106}Ru. Within the gastrointestinal tract, the colon will usually receive the largest doses because the residence times are greatest there. Ingesting of insoluble radionuclides that are alpha-emitters (such as plutonium) can result in low radiation doses since the alpha dose to tissues in the wall of the bowel is only a very small fraction of the dose to its contents.

The largest dose will be to organs that accumulate and retain the radionuclide. However, the variability in absorption of the ingested radionuclide in the gastrointestinal tract is responsible for the greatest uncertainty in the potential dose. Because radiation guidelines are usually conservative, it is likely that the commonly used absorption factors overestimate the amount of the radionuclide that is absorbed and hence the organ dose.

POTENTIAL CONSEQUENCES OF RADIATION EXPOSURE

The biologic effects of ionizing radiation are better understood than are those of exposure to any other potentially harmful element or compound. In large doses, ionizing radiation clearly causes cancer in humans, but fortunately radiation is a relatively weak carcinogen. Earlier in this report it is stated that no measurable increase in the risk of cancer has been observed in the Japanese population exposed to doses below 0.2 Gy (20 rad). Traditionally, radiation protection guidelines are predicated on a linear dose response, which assumes that the harmful effects of radiation are linearly related to the dose and that there is no threshold dose. Most experts believe this assumption is conservative; that is, it overestimates the effects of ionizing radiation at low doses because it ignores the potentially beneficial effects of the body's repair mechanisms. In the past, this probable overestimation of the risk was regarded as a good thing consistent with the still widespread philosophy that it is better to be safe than sorry. This philosophy holds true only when unlimited resources are available to protect the public health and the environment. Once resources are acknowledged to be limited, overestimates of a particular risk are ultimately harmful to the public health because funds are diverted from larger risks to protect society from smaller risks. This diversion of funds ultimately will result in greater mortality than would have occurred if resources were spent in proportion to the amount of health benefit that would be achieved.

To illustrate the magnitude of the cost of a dose reconstruction project, the U.S. Defense Nuclear Agency in the Nuclear Test Personnel Review (NTPR) over 10 years ago undertook the task of developing radiation dose estimates for about 203,000 military personnel and civilian employees of the U.S. Department of Defense who participated in the atmospheric nuclear weapons tests between 1945 and 1962. Approximately $100 million was required to estimate the doses received by the service personnel. The National Research Council was asked to review the scientific aspects of the NTPR project and in its report suggested that it is far more important to be able to state with a high degree of confidence that the dose is below some selected value than to estimate what the dose is when it is far lower (NRC 1985). In other words, it might be reasonable to establish an effective dose commitment over so many years below which it makes little sense to waste huge sums of money for further quantification.

To estimate health risks reliably, the most accurate estimate of dose to members of the public should be obtained. Estimates of the health risk of radiation exposure must be provided to the public in the most understandable terms possible, using a sound methodology and proper quality assurance. This is not simple because risk can be expressed in many ways. It can be stated in relative terms, such as the ratio of the risk in one population (or exposure group) to another or as the *excess* relative risk (the difference between the observed relative risk and 1, where 1 is the value expected in the absence of an effect). Risk also can be couched in absolute terms as the excess number of occurrences of an effect (for example, cancer deaths or cases) above the number "normally expected"— in the absence of exposure to other unnatural causes, with age and sex distribution and the length of observation taken into account—in the population of interest. "Normally expected" in this context would include exposure to ionizing radiation emanating from the Earth's crust, originating in outer space, or incurred medically. Risk also can be expressed as years of life lost, or as attributable risk, or as the proportion of cases that would not occur if the exposure had not happened.

It is not likely that any single measure of risk captures all of the information a member of the public might desire, especially when risk can vary with factors such as age, gender, and time after exposure. Because it is proportional to the spontaneous occurrence of the effect of interest (cancer, for instance), relative risk has limited utility in the absence of knowledge about the spontaneous or baseline rate; with relative risk alone the number of events that might occur cannot be predicted. Risk expressed as excess occurrences of a health effect (deaths or cases) has an immediacy that is readily grasped, but it depends on the spontaneous rate of such events and on the accuracy and completeness of their recognition. The rate of spontaneous cancers and the accuracy of diagnosis differ within a country from region to region, even from city to city, and neither can be estimated with assurance in small populations. These differences in the expression of risk become important in the present context, however, only when one seeks to extrapolate from the available partial-lifetime data to the full lifetime of a population such as that in the vicinity of the Hanford Nuclear Site, or when one seeks to extrapolate risks derived from one population to another one that has very different baseline rates for particular cancers. In the first instance, for example, the choice between the multiplicative or the additive risk projection model can lead to projections that may differ severalfold. In the second instance, the use of the Japanese A-bomb risk estimates can lead to substantial differences in the projection of site-specific risks where the baseline values are very different such as for cancers of the stomach, colon, liver and breast.

The following sections describe in general terms the approaches that can be used to assess potential risk that attends a given exposure. After that is a section on the types of uncertainty that can be encountered for such a procedure. The dose assessments have been divided into categories: preliminary, comprehensive, and individual. The preliminary approach is essentially a scoping process, the results of which can indicate the need for a more comprehensive dose assessment. If a dose assessment for a specific individual is desired, the process requires much more detailed information about that individual, including weight, height, lifestyle, and the like.

PRELIMINARY DOSE ASSESSMENT

The purpose of a preliminary dose assessment is to determine the need for a study of the health effects resulting from the exposure of a population to radiation. A preliminary assessment will normally precede a comprehensive health effects study, but when there is evidence of epidemiologic effects or widespread public concern, the health study can be initiated in lieu of, or in parallel with, the preliminary dose assessment. The features of a preliminary dose assessment can be subdivided into those related to input, method, and output.

An accurate estimate of the source term, with time dependence where necessary, is required. Spatial and temporal environmental data are needed, both from modeling studies and from measurement programs (modern and historic). Lifestyle and dietary data must be assumed, either from generic sources or from more local sources if available; individual data are not usually necessary at this stage. Depending on the pathways of importance, additional data can be required, to include food distribution and production data, population movements, and countermeasure data in the case of a specific accidental releases. Existing data on intake and external dose pathways will normally be used. Depending on the nature of the assessment, data on doses from other sources (natural, medical) and local background disease incidence might be needed.

This early assessment can help to focus further studies. For example, it might be sufficient to consider only individuals of a specific age, or only a limited number of organs, or a specific set of radionuclides. The principal intention of a preliminary assessment is to estimate the doses, and hence the risk of health effects, to an exposed population. For prolonged exposures (over several years), a useful approach can be to estimate the doses received by representative (but hypothetical) individuals. Their estimates can be extended to more extreme groups in the population by considering the sensitivity of the results to the assumption of unusual habits that can give rise either to greater exposure from standard pathways or to exposure from unusual pathways. The sensitivity of the results to variations in all of the parameters thought to be uncertain should be considered. A sensitivity analysis will generally be sufficient for the preliminary assessment, and a full uncertainty analysis should not be required (see Glossary for the distinction between sensitivity and uncertainty analyses).

The results to be obtained from a preliminary dose assessment will depend largely on the nature of the specific assessment and on the questions the study is designed to answer. In general, it is more useful to calculate annual dose rates and risks to reference individuals than to calculate risks based on committed effective doses. In addition, the overall dose to and risk of each cohort can be calculated by combining the representative results with the size and age structure of each cohort. With total cohort risk, the estimated total number of health effects can be obtained. The preliminary dose assessment should include a review of any bias that is likely to damage the successful outcome of a more comprehensive study. A related output will be an estimate of the statistical power of the comprehensive study.

COMPREHENSIVE DOSE ASSESSMENT

In a preliminary assessment, some releases and pathways will be found to be potentially more important than others. A more comprehensive study of the releases and pathways will be needed if an epidemiologic study is warranted. A comprehensive dose assessment also will be needed to provide individual specific doses that will reduce uncertainty in the dose estimate from the preliminary assessment. The preferred dose estimate from a comprehensive assessment is the annual absorbed dose to the target organs of the body.

For radiation exposures from external sources in the environment, the absorbed dose to body organs increases with decreasing body size; this effect is most pronounced at low photon energies and for absorbed dose to organs located near the middle of the body that are shielded by overlying tissue. Petoussi and others (1991) indicate that, at photon energies greater than 100 keV, absorbed doses in an infant can be about 40% higher than those in an adult male for exposures both from a contaminated ground surface and from submersion in a cloud source. Below 100 keV, the difference could approach a factor of 3 for deeper organs such as the ovaries and colon (Eckerman and Ryman 1993).

If the radiation exposure is from inhalation or ingestion, then physiologically based pharmacokinetic (PBPK) models will need to be developed to supplement the ICRP Publication 30 models (ICRP 1979) developed for a "reference" man for general applicability to radionuclides of various resident times in different organs of the body. Hence, PBPK models, which can be combined with the age-specific mathematical phantoms, can be used to make realistic calculations of absorbed dose as a function of age at the time of exposure and at a specific time after exposure. For radionuclides in bone, the absorbed dose can be delivered for many years after the initial exposure. Thus, intake could be considered daily or weekly and the absorbed dose calculated annually over the lifetime of the individual. As with the preliminary dose assessment, the absorbed doses due to low-LET radiations (x- or gamma rays, beta particles) should not be added to absorbed doses from high-LET radiations (usually alpha particles). The doses from the radiations with different LET values should be listed separately before computing "equivalent doses." The organ dose is especially important when developing doses to compare to site-specific health effects. However, there are circumstances when it is desirable to compare irradiations with different distributions among body tissues or to combine doses from non-uniform irradiation of various organs and tissues. Therefore, it may be necessary to calculate effective doses in addition to organ doses.

The applicability of absorbed doses to a particular situation is often influenced by exposure conditions that differ from the assumptions made in a model and by other factors that might alter these conditions at the location of the exposed individual. Any such assumptions must be justified and validated. For example, the radiation exposures inside a residence could be substantially different from those outside because of shielding, in the case of a contaminated ground plane, or because of filtration, in the case of inhalation of contaminated air. Environmental factors also can influence the time-integrated activity that characterizes exposures, and an individual's lifestyle can influence the extent of contact with radionuclides in the environment. For example, the time-integrated activity of a radionuclide in an urban environment can be substantially different from that in a rural area.

INDIVIDUAL DOSE ASSESSMENT

An individual dose assessment can be used in epidemiologic studies, and it will be needed for persons in the potentially exposed population who are interested in their own exposures. The preferred dose estimate from an individual risk assessment is the annual absorbed organ dose.

As with the comprehensive dose assessment, the radiation dose from internal sources as a result of either inhalation or ingestion can use PBPK models. The absorbed doses also must be calculated in terms of the LET of the various particles emitted (the high-LET alpha particles and the low-LET beta particles and photons).

The absorbed dose to an exposed individual is influenced by the exposure conditions for that person. As with the comprehensive assessment, environmental factors will influence the time-integrated activity that characterizes exposures, and an individual's lifestyle can influence the extent of contact with radionuclides in the environment. The dependence of the absorbed dose on other factors such as gender, lifestyle, and diet must be considered, and the details applied will be commensurate with the level of detail in the source term and the pathway analysis (Napier 1992). For example, the amount of time spent outdoors could affect the degree of exposure through inhalation or contact with radioactive materials deposited on soil or vegetation; the amount of milk consumed could affect the intake of radioactive iodines. It is unlikely that efforts to assess doses individually will be rewarding unless fairly detailed information is available on those factors that influence the individual dose.

UNCERTAINTY

Dose reconstruction involves the use of measurements and calculation procedures and scientific judgments made on the basis of available data. Each should be characterized by statements of uncertainty, providing measures of precision and accuracy. Statements of uncertainty must be made in defining the problem and presenting the results of investigations. One must state the degree of certainty with which one needs to know the answer, the sensitivity of the methods employed, and how the results of the studies are expressed. Statements of uncertainty must be expressed in probabilistic terms.

A quantitative estimate of dose uncertainty is important in determining the needed sample size, the achievable precision and statistical power, and the number of required measurements. The feasibility and utility of a study would depend on these considerations. In assessing exposure and absorbed dose, uncertainty should be expressed for physical, biologic, and computational methods. The calculations of uncertainty should be propagated throughout all calculations. Errors associated with physical measurements are likely to be smaller than are errors associated with biologic measurements, because the major contributor to the latter is interpersonal variability. In obtaining measures of propagated errors, procedures for incorporating methods of assessment of uncertainty for physical and biologic results are required.

It is helpful to separate uncertainty of knowledge of the state of nature from the act of making decisions. It is often necessary to make decisions even in the face of the uncertainty. This problem is solved by using decision thresholds above which action will be taken. That is, before deciding to carry out a study, decision makers need first to determine whether the uncertainty that will attend it is acceptable.

The dependence of the effects of radiation on dose rate and latent period makes it necessary to take into account the time dependence of doses delivered to the whole body or to specific organs. For instance, previous annual doses are used to assess the risk to individuals in the future. Accumulation of these doses over time will depend on factors such as residence history and yearly dose rates.

Direct measurement of radionuclide content in the body and in environmental samples, and individual dose measurements (physical or biodosimetric) should be used for validation of model predictions wherever available. External doses for different groups within a population can be evaluated from dose rate measurements in open areas or from radioactive contamination as determined by sample counting. Internal dose from inhalation and ingestion can be evaluated from radionuclide concentration in air and in food products. If the result of a direct measurement is different from a model prediction, preference must be given to the measured result.

SUMMARY AND RECOMMENDATIONS

Three levels of dose assessment can be envisaged: a preliminary assessment, a comprehensive assessment, and an individual assessment.

The purpose of a preliminary dose assessment is to determine the need for a full-fledged study of the health effects that results from the exposure of a population to ionizing radiation. The more comprehensive dose assessment will be needed if an epidemiologic study is undertaken. Among the aims of a comprehensive assessment are to provide the information necessary to compute individual specific doses and to reduce the uncertainty in the dose estimate arising from a preliminary assessment. Finally, individual dose assessments define doses to particular individuals in the potentially exposed population who may be interested in an evaluation of their potential health risk. The most useful dose quantity in a preliminary dose assessment is the effective dose; in a comprehensive dose assessment or in the computation of an individual dose, the preferred dose estimate will be the annual absorbed organ doses from low-LET and high-LET radiations. However, it should be noted that published intake dose conversion factors are for committed doses. Hence, except for the equilibrium situation with short effective half-life radionuclides, it is more difficult to calculate annual doses for various years following intake.

The committee makes several recommendations as follows:

18. All exposures from external sources, inhaled radionuclides, and ingested radionuclides should be considered; when certain pathways or other factors suggest that a particular source term or radionuclide will not contribute substantial dose, reports should explain why these sources or specific radionuclides were not considered in the final estimations.

19. Dose assessment should proceed at three levels: preliminary, comprehensive, and individual dose assessment.

20. Acceptable levels of uncertainty should be defined before a decision is made to carry out a detailed study.

21. A readily available set of intake-to-annual dose conversion factors for long-lived radionuclides should be established.

22. Doses should be expressed as effective doses in a preliminary dose assessment and objective criteria should be used to decide whether it is warranted to embark on a full-fledged study.

23. Dose estimates in a comprehensive dose assessment should be expressed as the annual organ absorbed doses from low-LET and high-LET radiations. Estimation of the effective dose may also be helpful.

24. In an individual dose assessment, the doses should be described separately as the annual organ-absorbed doses from low-LET and high-LET radiations.

6

Biologic Dosimetry and Biologic Markers

THE TERM BIOLOGIC MARKER encompasses many biologic end points. For mutagens or carcinogens, these end points can be characterized according to where they occur in the process leading to an environmentally mediated disease, such as cancer.

In order to produce a biologic marker, a harmful agent must interact with cellular macromolecules, including DNA, to induce some chromosome- or DNA-damaging event, such as an aberration, the formation of a micronucleus, a DNA adduct, or a somatic mutation. Biologic markers such as aberrations, micronuclei, DNA adducts, or mutations can be measured in a surrogate tissue or in the specific tissue of interest, that is, the target tissue for the effect. Chromosome- or DNA-damaging events also can be measured, using molecular techniques, in oncogenes and suppressor genes. These end points overlap with manifestations of early frank pathology, such as altered cell structure or function, metaplasia or dysplasia, early in situ carcinoma, and, finally, cancer. A sequential ordering of markers is useful because it can define the stage of pathogenesis by end point. Damage evident in the cancer genes themselves is really more a molecular manifestation of disease than it is a biologic marker.

There are three kinds of biologic markers: markers of exposure or dose, markers of effect, and markers of susceptibility. Biologic markers of effect record biologic responses in individuals who have been exposed to a genotoxic agent, but markers of exposure (or dose) do not necessarily indicate effects. Superimposed on this are markers of susceptibility; those that could be used to identify persons who are at increased risk of devel-

51

oping a disease that could be triggered by a given exposure. Included here might be persons whose ability to repair DNA damage is limited.

Biologic effects seen after moderate and low doses of ionizing radiation are almost invariably the result of damage to the genetic apparatus. Because of the centrality of the genetic material to the induction of damage, and because it is possible to visualize damage by molecular genetic or cytogenetic means, methods that use markers of effect after exposure to low to moderate doses of radiation have concentrated on somatic genetic effects.

MARKERS OF EXPOSURE AND DOSE

Cytogenetic Markers

A chromosome aberration occurs when cells are irradiated and the chromosomes are broken and can rejoin with time after exposure. The kinetics of induction and repair have been carefully studied. For acute exposure to low linear energy transfer (LET) ionizing radiation there is a linear dose-response relationship for simple terminal deletions. Aberrations that require two independent breaks and the interaction between two chromosomes increase linearly at low doses and as the square of the radiation dose at higher doses. When the dose is from high-LET radiation or if the low-LET radiation is protracted or fractionated, aberrations increase as a linear function of dose. Because relevant calibration curves for aberrations can be obtained using human lymphocytes in vitro, it is possible to use the frequency of aberrations measured in lymphocytes of the exposed individual to estimate radiation dose when actual physical measures of dose to an individual are unavailable. Results of studies of people accidentally exposed to high radiation doses indicate that doses estimated from yields of aberrations coincide well with the measured doses (Gooch and others 1964, Bender and Gooch 1967).

Chromosome aberrations induced in G_0 human lymphocytes have been the system of choice for a biologic dosimeter used to quantify the dose to which an individual has been exposed or to verify or corroborate a suspected exposure for which no physical dose measurements have been available (Dolphin and others 1973, Lloyd and others 1987). These studies used mainly a dicentric aberration, an unstable aberration whose frequency decreases with time after exposure. This frequency depends on cell turnover rate; aberration persistence can be relatively long in nonproliferating cells. The background concentration of dicentric cells in unirradiated persons is low, as little as 1-2 dicentrics per 1,000 cells in T-lymphocytes (Littlefield and others 1990), and there is little variability among individuals, so that small radiation-induced increases can be quantified. This system can be used to estimate doses as low as 0.10 Gy (10 rad). Below that, the sample size required for statistically reliable results is so large it is impractical to obtain (Evans and others 1979; Lloyd and others 1980).

Classic cytogenetic techniques for estimating a dose can be used not only for measurement of a person's exposure, but also for limited epidemiologic purposes if one keeps in mind the unstable nature of aberrations such as dicentrics and the fact that loss may vary between individuals with time after exposure (Littlefield and others 1990). However, if one wants to estimate the dose to which a population has been exposed, as the dose becomes lower than 0.10 Gy (10 rad) the method becomes labor intensive and impractical. If there is a question about the magnitude of the exposure, such an evaluation can establish upper bounds of exposure and help define the dose to populations. It must be kept in mind that the decay found in this end point makes it useful only for an individual recently exposed to radiation (Wolff 1991). Authors have reported values for the average disappearance half-time of lymphocytes containing dicentric and centric rings ranging from 130 days (Ramalho and Nascimento 1991) to 3 years (Lloyd and others 1980).

Evaluation of the frequency of stable chromosome aberrations (those that do not decrease with time) has been made possible by techniques that measure translocations between chromosomes. This is done by evaluating banded chromosome preparations or by using a less accurate but more rapid size-grouping method. Such techniques were useful in measuring aberrations in the survivors of the atomic bombing of Hiroshima and Nagasaki at long times after radiation exposure.

Fluorescent in situ hybridization (FISH) can be used to further define stable chromosome aberrations (Pinkel and others 1988). It is a promising cytogenetic method for determining the dose of radiation to an individual, and consequently, to a population, particularly for those receiving protracted exposures or for those exposed a long time ago (Straume and Lucas 1995). The technique uses chromosome-specific fluorescent probes to "paint" specific chromosomes so that exchanges involving that chromosome can be identified rapidly (Lucas and others 1989a; Gray and others 1991). When combined with further development of DNA probes specific for centromeres (Lucas and others 1992a), this technology can now efficiently and accurately detect reciprocal translocations in human cells. In addition, full genomic translocation frequencies can be obtained from FISH translocation frequencies after staining only a small fraction of the genome (Lucas and others 1989b). This finding permits scaling to the full genome from only a few painted chromosomes. Fluorescent in situ hybridization has been used to demonstrate that dicentrics and reciprocal translocations are induced with identical frequency (Straume and Lucas

1993; Nakano 1993), making this technique as sensitive as the use of dicentrics. The sensitivity of this method depends on the background level, the ability to score large numbers of cells rapidly, and the ability to detect even small translocations. Theoretically, it could detect lower doses than is possible using conventional cytogenetic methods. Additional measurements of the background number of translocations as a function of age, stability with time after exposure, and the amount of variability from one person to another are needed before the full promise of this assay can be realized. Some limited preliminary information exists about using fluorescent in situ hybridization to determine reciprocal translocations in unexposed persons. This suggests that background frequencies are in the range of 4-8 per 1,000 lymphocytes and that they increase with age, possibly because of their persistence (Lucas and others 1992a, 1994).

Another cytogenetic end point that has been used to estimate exposures is the measurement of micronuclei in populations of exposed cells. Micronuclei are formed when cells with broken chromosomes divide and the acentric pieces do not proceed to the poles at anaphase. These fragments, or lagging chromosomes, are not included in the daughter nuclei and they form recognizable, diminutive chromatin bodies or micronuclei in the daughter cells. The evaluation of micronuclei is much easier to perform than is chromosome analysis. It does require that the cells divide after the insult for the expression of micronuclei. To ensure that only dividing cells are scored, cells are treated with cytochalasin B, which blocks cytokinesis and results in binucleated cells. Only the binucleated cells are evaluated for the formation of micronuclei. The problems with this procedure are that at very high doses there is evidence that micronuclei can join to decrease frequency and that there is a high and variable background frequency in human lymphocytes. It has been estimated that the high variability and the high background limit the sensitivity to doses of radiation of 0.3 Gy (30 rad) or more (Prosser and others 1989). The rapid nature of the test, the possibility of automation, and the response to multiple environmental insults do make this useful for identifying markers of exposure for some experimental questions and exposure conditions.

Genetic or Molecular Markers

One method of detecting exposure uses fluorescent-labeled monoclonal antibodies to detect mutations or losses of the alleles on chromosome 4 that code for glycophorin A, a glycoprotein responsible for the M and N blood types found on nonnucleated erythrocytes (Langlois and others 1993; Grant and Bigbee 1993). Because the gene has no known function, it is now thought to be neutral, and thus free from biases caused by selection. The technique uses flow cytometry to separate the cells, and its use is limited to the half of the population that is heterozygous MN. Erythrocytes that have lost either the M or the N allele, and are thus M0 or N0, can be detected, as can cells that are homozygous MM or NN (from somatic recombination). The M0 and N0 variants increase with radiation dose, and this assay could be used as an marker of exposure or dose for individuals. The observation that the grouped data from the survivors of the atomic bombing of Hiroshima and Nagasaki showed a dose-related increase in M0 and N0 variants many years after the exposure indicates that the phenomenon has an extremely long half-life. A positive correlation has been found between increases in M0 variant cells and chromosome aberrations in the survivors many years after the exposure (Kyoizumi and others 1989). Both of these observations suggest that M and N alleles are markers of exposure if the population size is large enough and if the dose is high enough. Because there is no in vitro calibration that can be used to determine what response is expected from exposure to radiation, the assay has not been validated and further work is needed.

One of the most extensively studied human mutation systems involves the induction of mutations in the X-linked hypoxanthine phosphoribosyl transferase (*hprt*) locus in human cells. This locus is involved in the salvage pathway whereby hypoxanthine or guanine is phosphorylated so that it can be incorporated into DNA. Loss of the locus or its gene activity conveys resistance to the purine 6-thioguanine, which kills wild-type cells. Detection of cells that will grow to form colonies in the presence of 6-thioguanine allows analysis of the molecular nature underlying the mutations (Albertini and others 1982; Morley and others 1983). In vitro experiments have demonstrated a linear increase of *hprt* mutants as a function of dose, which suggests that *hprt* could be useful for biologic dosimetry. However, data on *hprt* mutants in survivors of the atomic bombing of Hiroshima and Nagasaki show extreme scatter and decreased yield in people several years after their exposure, suggesting that a loss of mutants could have occurred by negative selection. A further complication exists with the use of such loci that are detected by selection, in that these mutants could represent a cluster that has been expressly selected. This could inflate quantification of the effects of the inducing agent (Nicklas and others 1986). As more is learned about the quantitative relationships of in vivo induction of *hprt* mutations in humans their use for biodosimetry could be validated.

Other gene loci have been tested for utility as biologic markers of dose. Among those is the autosomal HLA-A locus. Wild-type human T-lymphocytes are killed by exposure to complement plus monoclonal antibodies against HLA-A2 or HLA-A3. Mutant cells that do not express the surface antigens survive (Janatipour and others 1988). Mutants can be induced by radiation, but the data that show such effects have been obtained only at high doses. Akiyama and colleagues (1992) studied survivors in Hiroshima and Nagasaki and failed to find a radiation effect at this locus. Because of the loss of mutants with time the technique will not be a useful lifetime dosimeter. The absence of any information regarding the half-life of the mutants and the reasons for their loss with time makes its usefulness as a biologic marker of dose problematic.

Lymphocyte T-cell antigen receptor (*tcr*) assays also have been investigated (Akiyama and others 1992). *Tcr* alpha and *tcr* beta chains form a heterodimer involved in the surface expression of CD3 complexes. If a mutation occurs in either of these, the CD3 complex cannot be expressed on the cell surface. Such mutants are detected as CD3⁻ cells as determined by the use of monoclonal CD3 antibodies. When Hiroshima and Nagasaki survivors were tested, no increase in mutants was found over the entire dose range, although female patients with cancer of the reproductive system who received very high doses of radiation showed a radiation-induced increase in mutants at this locus. The mutant frequency declined with a half-life of about 2 years, which might have contributed to the lack of an effect observed in the survivors. It should be noted, however, that the standard cytogenetic study of unstable chromosome dicentrics still shows an increase with dose in the survivors, even though that end point has a half-life of anywhere from 2 to 5 years. It thus appears that in the absence of a known dose-response curve, including effects at low doses, and in the absence of studies indicating why the effect was negative in the survivors, effects measured with this locus are not as promising as are those that use the *hprt* locus.

Attempts also have been made to study mutations at the beta-globin locus. The method used automated techniques to pick up rare red blood cells that contain mutant hemoglobin. Mitigating against the use of beta-globin as a marker of dose for radiation is the fact that the method of assay is extremely cumbersome and expensive, that the mutations observed are very rare, and that the mutations consist of only a single base change—they are true point mutations, which are rarely induced by ionizing radiation.

The committee recognizes that if biologic markers of dose are to fulfill their promise more research will be required to validate the response of specific markers to radiation dose, to identify new markers, and to better characterize the limitations and sensitivities of known markers.

Combined Biologic-Marker Assays

Perhaps the most useful biodosimetry analysis that can be performed on people exposed to low doses of ionizing radiation is to measure multiple end points using several of the various assays simultaneously. This provides complementary information from each of the assays, such as persistence of stem cell effects from the glycophorin A assay, mutation spectra from the *hprt* assay, and good dose sensitivity from the chromosome aberration assays. In addition, each assay measures a somewhat different kind of genotoxic effect caused by radiation exposure, such as clastogenicity (chromosome aberrations) as compared with loss of enzyme activity (*hprt*) or loss of allelic expression (glycophorin A). Confounders would be somewhat different for each assay, so the precision of dose reconstruction and the prediction of subsequent health effects from radiation would be more precisely defined. The disadvantages of this approach are its cost and the need for a sufficiently large blood sample from each person in a large population.

The committee suggests that when biologic markers are used to assess dose, multiple end points should be measured using various assays simultaneously so that the abilities and value of the assays can be maximized.

MARKERS OF EFFECT

Occasionally, the same end point can be a marker of both dose and effect, depending on the type of damage and the time when the marker is studied. For example, radiation exposure can produce a well-characterized number of chromosome aberrations per unit of dose, and aberrations are good markers of exposure during the early stages of cell proliferation in tissue. As the cells divide, form hyperplastic nodules, progress to benign neoplasms, and finally form radiation-induced malignancies, the cells lose unstable aberrations but might retain stable chromosome aberrations or other genetic lesions that survive cell division and can be classified as markers of disease or effect. It is essential to conduct studies that will relate markers of dose to markers of effect and explain how both relate to the induction of a specific disease or to the incidence of cancer.

There are unique chromosome aberrations that are characteristic of defined cancer types. For example, the Philadelphia chromosome is a biologic marker for chronic myelogenous leukemia. The presence of this marker in bone marrow cells predicts the development of the disease. There are other examples for other cancers, such as the loss of part of chromosome 17 during the induction of colon cancer and the rearrangement in either chromosome 1 or 16 or both during the development of breast cancer. These genetic changes are biologic markers of effect. Other molecular markers suggest increased risk for the development of radiation-induced cancer. For example, individuals who are heterozygous for the Rb gene (retinoblastoma) are at an increased risk for both spontaneous retinoblastoma and radiation-induced cancer of the skeleton.

MARKERS OF SUSCEPTIBILITY

Heterogeneity with respect to responses to insults like radiation raises concern over potentially susceptible subsets of human populations. Susceptible subpopulations would contain individuals that have different levels of protection from or sensitivity to the genotoxic effects of ionizing radiation. Susceptible groups, such as persons with defective DNA repair (ataxia telangiectasia) or those who are heterozygous for the Rb gene and are sensitive to radiation-induced cancer, are known to exist. Knowledge of the marker responses of such individuals is needed. Differences in response in a population could just as well result from differences in individual susceptibility as differences in individual exposure or dose. It must be asked whether there are specific biologic marker responses that could be identified in such populations. Such differences could lead to confusion in evaluating biologic markers of exposure and effect. More research is needed to identify the responses of susceptible subsets of human populations.

MARKERS IN RETROSPECTIVE DOSIMETRY

Which biologic markers are most appropriate for use in monitoring a human population? The answer will depend on a study's goals. Markers of exposure would be identified first where exposure assessment is the primary concern, with the proviso that the sensitivity of the marker is adequate to the task at hand. By contrast, marker effects would be studied if an exposure has resulted in identifiable genetic damage. Markers of susceptibility are needed to interpret inter-individual differences in response to radiation or genotoxic chemical exposure and, in some cases, to select those persons at greatest risk.

The important questions are whether biologic markers of exposure can be useful in dose reconstruction and, if so, how? For retrospective dose reconstruction, it is generally agreed that markers of exposure are not useful below an acute dose of 0.1 Gy (10 rad), at least with current technology. Some of the problems include the large degree of individual variability in background for individual markers. Also, there is the problem of decay or loss of some markers with time after exposure. This is probably less true for markers in stem cell lines than it is for those in short-lived cells.

The greatest difference between the use of biologic markers in future events or accidents and using them to reconstruct accidents that occurred in the past is that the kinetics of measuring schemes can be used to advantage. For example, among the most sensitive markers for radiation exposure is the enumeration of unstable chromosome aberrations. If this marker is measured within a year after an acute exposure there will be little decay and the sensitivity will allow it to serve as a good dosimeter. Of course, chronic exposure will induce an accumulation of unstable chromosomal aberrations, and the combination of new aberrations, with the decay of the older lesions, would confuse the evaluation of the total dose. In a similar manner, the sensitivity of *hprt* and *tcr* assays shows decay with time after exposure. The disadvantages associated with unstable markers are avoided when stable markers, such as reciprocal translocations measured by fluorescent in situ hybridization, are used.

One area in which the use of biologic markers could be important is in the identification of persons exposed to high doses of radiation. In contrast to concerns about evaluating low doses in large populations, biologic markers can effectively identify persons exposed to radiation levels above 0.1 Gy (10 rad). This is just as valuable a public health finding—to demonstrate whether specific persons did or did not receive a high dose—as is documenting the overall range of doses for epidemiologic purposes.

MARKERS IN EPIDEMIOLOGY

Biologic markers of effect could be useful as epidemiologic end points, although the validity of specific markers as surrogates or predictors of disease must be demonstrated. Knowledge of the relationships between specific markers and specific disease end points is required if the use of markers is to have epidemiologic value in an exposed populace. There must be more than a simple correlation between a marker and an effect in a population; that is, there must be studies that correlate the marker with the effect (such as cancer) in specific persons.

There are human populations, primarily those given therapy for cancers such as Hodgkin's disease, in which the secondary cancer (acute leukemia) rate is 7-8% after 7 years. Later comparison of biologic marker status in individuals who do or do not develop disease could be compared in a nested case-control study that could shed light on the relationship of the marker to the subsequent development of disease and whether the worker might be used in broader studies as a true surrogate for the disease. Therefore, the committee suggests careful consideration be given to creating a cryopreserved repository of tissue, blood, or both, from persons at high risk for cancer after a known exposure to radiation (or other agent).

Markers such as *hprt* have been used in cases of human exposure to chemicals including ethylene oxide, butadiene, and polycyclic aromatic hydrocarbons. In these cases, the markers were correlated either with DNA adducts or with urinary excretion products. The question asked in these cases was not whether the persons had been exposed—this was already known—but whether there was a demonstrable effect of the exposure. Even in these studies, with a demonstrated genotoxic effect, there could be no clear answer about the relationship of the marker and a specific disease end point.

SUMMARY AND RECOMMENDATIONS

The use of biologic markers for dose reconstruction or epidemiologic studies associated with dose reconstruction involves extensive effort and expense, and many of the techniques for finding markers carry much uncertainty (Albertini and others 1990; Grant and Bigbee 1993; Grant and Jensen 1993). Background variability is a major problem with markers other than dicentrics measured by cytogenetic methods. A promising advance is in the scoring of stable chromosome translocations using fluorescent in situ hybridization. Validation measurements made by fluorescent in situ hybridization have shown that the frequency of reciprocal translocations in whole-body-exposed individuals is constant with time after exposure. The validations have provided dose reconstruction results that are consistent with independent dosimetry methods (Straume and others 1992; Lucas and others 1992b, 1993, 1994). The glycophorin A assay also detects a long-lived change in M and N blood cells that can be related to the dose of radiation and that might be useful as a test for a marker of exposure. The use of multiple assays could help to reduce the uncertainties that arise from interindividual and intraindi-vidual variability. In any case, the most sensitive methods can reliably detect only those markers that indicate acute doses greater than 0.10 Gy (10 rad). At lower doses and dose rates, the use of currently known markers is unlikely to help with dose reconstruction. For additional information in this area, the reader is referred to a recent review of this subject by Mendelsohn and colleagues (*Biomarkers and Occupational Health: Progress and Perspectives, 1995*).

The committee makes the following four recommendations:

25. For biologic markers to be useful in dose reconstruction, research will be necessary to
- **measure the stability of persistent biologic markers,**
- **define "calibration curves" for low to moderate and chronic exposures,**
- **determine the frequency of specific markers in unexposed populations,**

• define the sources of inter-individual variability for various markers, and

• develop better definitions of marker responses after partial-body (external) or specific organ (internal emitter) exposure.

26. New assays should be developed to address the problems with individual variability in background, with identification of differences in individual susceptibility to radiation genotoxicity, and with the lack of sensitivity for quantifying low radiation exposures so that acute doses greater than 0.1 Gy (10 rad) can be reconstructed.

27. Biologic markers of effect should be used as epidemiologic end points. However, until clear connections are established between the marker and the disease, their use could be misleading rather than illuminating.

28. As the utility of biologic markers becomes established and accepted, the committee recommends that the CDC develop procedural strategies for conducting field studies for both specimen collection and laboratory analyses in the event of an acute release of activity.

7

Epidemiologic Considerations

EPIDEMIOLOGIC STUDIES OF populations potentially exposed to ionizing radiation as a result of the release of radionuclides into the environment can be of two general forms, monitoring or formal, and serve several possible purposes. Monitoring studies inform members of potentially affected population groups of the nature and magnitude of the risks that might have been imposed on them. Monitoring studies also can guide people who are responsible for the facilities to identify measures that must be taken to minimize any future risk to surrounding populations. Formal epidemiologic studies can increase scientific knowledge about the quantitative risk that attends exposure.

Although, in principle, much could be learned from experience, for example, with the accident at Chernobyl in the Ukraine, it must be emphasized that deriving new scientific knowledge from studies of exposed populations around nuclear facilities is difficult. Radiation doses received by the latter individuals probably cannot be quantified precisely and the number of radiation-induced illnesses generally is smaller than the number one would expect to find from all causes in that population. Carefully done epidemiologic studies can inform those who might have been affected about the approximate magnitude of the risk, which diseases (chiefly, specific forms of cancer) they and their physicians should anticipate, and what amount of health monitoring is appropriate.

The design of a study and the data needed will depend on the study's aims. Dosimetric data are essential for any epidemiologic study, but the detail and accuracy needed depend on the purposes to be served. If the need is for a monitoring or scoping study, then general information about doses will suffice; a study that is expected to contribute to scientific information about quantitative radiation risk requires careful individual dose estimates.

Just as it would be inappropriate to begin an epidemiologic study without knowledge of the radiation doses involved, so it is ineffective to begin a detailed study of the exposures and doses before knowing what kind of epidemiologic study will be undertaken. Each kind of study will affect the other. Comprehensive dosimetric studies should not be undertaken before the full epidemiologic study is designed unless, of course, it is found in a preliminary dose assessment that exposures are too low or the potentially affected population is too small to make an epidemiologic study statistically worthwhile. As a rule, dosimetric and epidemiologic scoping (screening) studies should be undertaken in parallel, so that, at the conclusion of the dosimetric scoping study, an informed estimate of the expected magnitude of risk and the statistical power of a potential epidemiologic study can be derived.

This chapter considers various aspects of epidemiologic studies in the context of dose reconstruction, including strengths and limitations, interactions between epidemiology and dose reconstruction, and study design and methods.

QUANTITATIVE RISK ASSESSMENT: STRENGTHS AND LIMITATIONS OF EPIDEMIOLOGIC STUDIES

Knowledge of the biomedical effects of exposure to ionizing radiation has expanded enormously since the end of World War II. The research has ranged from the exploration of basic cellular mechanisms to extensive epidemiologic studies of exposed populations, including those exposed occupationally (Kendall and others 1992, Gilbert and others 1993), those exposed to therapeutic and diagnostic medical sources (Boice and Land 1982, NRC 1990), and those exposed to nuclear weapons. Prominent among the last group are the studies of the survivors of the atomic bombing of Japan (Shimizu and others 1990), including those summarized in a special supplement in *Radiation Research* (Mabuchi and others 1994, Thompson and others 1994, Preston and others 1994, Ron and others 1994).

As a consequence of the availability of this large body of knowledge and the need to set radiation standards for public and occupational exposures, radiation protection groups have continuously reviewed the literature to estimate the risk that results from radiation exposure. The greatest interest has been in the risk associated with exposure at low doses and dose rates. The most recent efforts to provide risk estimates are described in reports of the International Commission on Radiological Protection (ICRP, 1991), the National Research Council's Committee on the Biologic Effects of Ionizing Radiation (NRC 1990), and the United Nations Scientific Committee on the Effects of Atomic Radiation (UNSCEAR 1988, 1993, 1994).

Because estimates of risk that are based on direct study of persons exposed at low doses have often been too imprecise to be useful, risk estimates have been obtained primarily through extrapolation of data from studies of persons exposed to moderate or high doses and at high dose rates. Most scientists are reasonably confident—because of experimental evidence—that linear extrapolation is not likely to underestimate risk but there is appreciable uncertainty about the risk that results from exposure at low doses and dose rates (NCRP 1980). Because the Japanese atomic bomb survivors have been so important in radiation risk assessment, uncertainty about extrapolation from the Japanese group to another population also must be considered.

Despite their limitations, numerical expressions of risk are useful in dose reconstruction and epidemiologic studies because they provide a means to estimate the expected number of health outcomes of a particular kind, such as cancer, and hence to evaluate the probable meaningfulness of an epidemiologic study. Several agencies, including the U.S. Nuclear Regulatory Commission (Gilbert 1991b), ICRP (ICRP 1991, Land and Sinclair 1991), and the U.S. Environmental Protection Agency (US EPA, 1994), have provided estimates of the risk of radiation-related fatal cancers that can be used in scoping or subsequent studies. These estimates are summarized in Table 7-1. The risks are expressed as excess fatal cancers per 10^4 person-gray for various tissue sites. The differences in risk depend on the different projection models used and differences in the interpretation and adjustment of the basic data derived from studies of the Japanese atomic bomb survivors and other highly exposed populations. It should be noted that the largest differences among risk estimates involve stomach, colon, lung, and breast cancers, and the most discrepant values are usually those of ICRP. These differences are not likely to be important in scoping studies, however, in which the primary objective is to provide a preliminary estimate for use in deciding whether further detailed studies are warranted.

As shown in Table 7-1, various interpretations can be made with the basic data on cancer site-specific mortality and the reporting of the estimates with the decimal place given implies a false precision. Since most of the data on the occurrence of excess cancers (with the exception of breast cancer) are derived from high-dose, high-dose-rate studies, the National Research Council and others recommend application of a dose, dose-rate effectiveness factor (DDREF) between 2 and 10. For low-dose,

TABLE 7-1 Estimates of the Expected Number of Excess Fatal Cancers in a Lifetime at Specific Sites Among 10,000 Individuals Exposed to 1 Gy or 100 Rad (Males and Females Combined, DDREF = 1)

Cancer Site	ICRP[a]	NIH[b]	EPA[c]	NRC[d]
Esophagus	16.2	21.3	18.1	14.9
Stomach	29.3	27.4	88.7	74.3
Colon	381.0	109.0	196.4	149.0
Liver	30.0	30.0	30.0	29.7
Lung	265.0	78.7	143.2	149.0
Bone	9.3	9.3	1.9	8.1
Skin	2.0	2.0	2.0	1.8
Breast	116.0	32.7	46.2	46.2
Ovary	47.5	25.0	33.2	32.2
Bladder	64.0	38.9	49.7	49.6
Kidney	]—[[e]	]—[	10.9	]—[
Thyroid	7.5	7.5	6.4	6.4
Leukemia	110.0	97.9	99.1	89.9
Residual	325.0	227.0	246.2	193.0
Total	1403	953	973	844

[a]International Commission on Radiological Protection (ICRP 1991).

[b]These calculations were made by Land and Sinclair in 1991, but are commonly referred to as the NIH model.

[c]U.S. Environmental Protection Agency (US EPA 1994).

[d]Nuclear Regulatory Commissison (Gilbert 1991b)

[e]]—[= no estimate.

low-dose-rate radiation, if a DDREF of 2 (the most conservative factor) is applied to the data in Table 7-1 for all tissue sites except the breast, the total lifetime risk of a fatal cancer after whole-body irradiation would be 5.1×10^{-2}/Gy (5.1×10^{-4}/rad). For total body effects based on the atomic bomb survivors, the BEIR V committee determined risk factors of 7.5×10^{-2}/Gy (7.5×10^{-4}/rad) for females and 7.9×10^{-2}/Gy (7.9×10^{-4}/rad) for males without application of a DDREF (NRC 1990). A range of risk factors can therefore be chosen for computing excess cancer risks from radiation exposures; however, most dose estimates encountered in dose reconstruction studies are usually for specific tissues (thyroid or bone, for example) and at low dose rates for cumulative total doses that would generally be considered as low. The lifetime radiation risk factors for dose reconstruction studies would generally range from 5 to 8×10^{-2}/Gy (5 to 8×10^{-4}/rad).

In evaluating whether a proposed epidemiologic study is likely to increase scientific knowledge of radiation effects, it is important to consider the extensive body of scientific data on radiation risks and the models for evaluating risks that have been developed and endorsed by the groups listed in the table. The use of models, together with the use of disease rates obtained from cancer registries and vital statistics, can provide a means of estimating the number of radiation-induced cancers that will occur in a particular population exposed to a particular dose. The age and sex distribution must be specified, and given this information, one can estimate both the number of cancers that could occur as a result of the exposure of interest and the number that would occur in the absence of the exposure. Once these values have been determined for various categories defined by dose and other factors, it is possible to calculate the precision with which a proposed study can estimate risk and the probability that the study can statistically detect an excess of a given magnitude. Usually, if the number of radiation-induced cancers is small both in absolute value and as a proportion of the non-radiation-induced cancers, it will not be possible to determine with confidence whether or not an excess number of cancers has occurred. In addition to assessing the potential of a study to evaluate risks based on the assumption that current models are correct, it also could be of interest to evaluate whether risks that are several times larger than those based on current models could be excluded.

A simple example is helpful: If 10,000 persons with an age and sex distribution typical of the United States as a whole are each exposed to 10 mGy (1 rad) and followed over their full lifetimes, use of the BEIR V-recommended risk model (NRC 1990) would lead to an expectation of about 7.5 radiation-induced fatal cancers in the group. One would expect to see about 1,800 fatal cancers in a group that has not been exposed to the additional radiation. The problem for the statistician is to determine whether, given the variability in cancer occurrence, an expected excess of 7 or 8 cancer deaths could be detected with a particular level of confidence. Would the study have the statistical power to detect such an excess? If the answer is yes, then the study might reasonably proceed, although other issues such as the feasibility of identifying the population and of ascertaining the causes of the deaths also would need to be considered. If the answer is no, then the merit of proceeding with a formal study is doubtful.

The principal strengths of the epidemiologic studies associated with dose reconstructions come from the fact that humans are the populations of interest. No extrapolations are required from animal to human, from high-dose studies, or from populations with other ethnic, lifestyle, or sociodemographic characteristics. Such studies avoid many of the uncertainties involved in extrapolations and maximize the potential to address public concerns. They also can provide direct information about sensitive subgroups of the population, and they can offer the chance to incorporate any biologic marker studies that seem warranted.

The value of a low-dose study, such as one done near a nuclear site, is increased if it is understood that the study results do not stand in isolation and that the study is performed in the context of adding to the body of information from other radiation studies. A systematic meta-analysis is often useful as a framework for integrating and comparing the results of several such studies (Greenland 1987), although joint analysis of the raw data from several studies is preferable when the data are available and comparable (Gilbert and others 1993).

Epidemiologic studies done in conjunction with dose reconstructions near nuclear sites or in the aftermath of accidents are subject to several limitations. The exposed population could be poorly defined, inhomogeneous, and transient. It could be relatively small, so that only large effects could be detected reliably. This problem is exacerbated when the diseases of interest are relatively rare and the risk per unit dose is expected to be small. The exposures could be difficult to evaluate because of uncertainties about exposure amount, duration, and timing. The diseases or medical conditions of interest could have long latencies and might not have been recorded reliably. For some outcomes, public concern could have created dose-dependent variations in disease surveillance, which could bias the study results. There also could be effects of chemical contaminants from the site and from other industrial sites nearby because of the lack of detailed exposure information. Epidemiologic studies also can be confounded by dietary or lifestyle factors (of which smoking habits are particularly important), especially when they deal with low-dose data.

EPIDEMIOLOGY AND DOSE RECONSTRUCTION

For a dose reconstruction to be useful for epidemiologic purposes, it must be designed to allow for the calculation of annual organ doses. Furthermore, an epidemiologic investigation often involves tradeoffs relating to bias and precision in the estimates of exposure and other information that can be crucial not only to the dose reconstruction study but to the epidemiologic one as well. It follows, then, that the involvement of an epidemiologist in the early phases of the dose reconstruction will ensure attention to the potential for bias in the reconstruction of individual doses. Epidemiologists need to understand the details of dose reconstruction both to collect the appropriate information on dietary and lifestyle factors that could significantly affect dose calculations and to evaluate in the epidemiologic analysis various uncertainties inherent in dose estimation.

Epidemiologic evaluation also can contribute to dose reconstruction. In particular, epidemiologic data could be used to validate models and assumptions used in dose reconstruction and to assess the usefulness of different dose estimates, starting from the simplest, such as distance from the site, to the most complex, such as those based on models of environmental pathways and on the use of biologic markers. While in theory epidemiologic observations can contribute to validation of dose estimates, as a practical matter they are not likely to do so except for large populations exposed to high doses.

As noted in the Introduction, if a preliminary dose assessment results in estimates of exposures that are of potential concern, delaying the epidemiologic investigation until a detailed dose reconstruction is complete could not only diminish the usefulness of the reconstruction but could jeopardize a warranted epidemiologic study. For example, the more time has elapsed from the period of interest the more difficult it usually becomes to assemble the complete population needed and to reconstruct the information required for exposure assessment. Moreover, as previously stated, the likelihood of introducing a variety of biases increases as the publicity surrounding dose reconstruction efforts broadens. Finally, negative public perception about delays in directly addressing the issues of concern could create difficulties and require additional efforts that cannot be justified on scientific grounds.

STUDY DESIGN

In the introduction to this chapter, two types of epidemiologic studies were distinguished: monitoring studies and formal epidemiologic studies. The decision about whether a monitoring study should be done will depend on the amount of information available about the approximate levels of exposure and the amount of concern about possible effects on health. To determine whether a formal epidemiologic study should be undertaken, it is necessary to evaluate the power of the proposed study (which will depend on the population size and the magnitude of the doses) and to evaluate the feasibility of obtaining the necessary data on health effects, exposure, and potential confounders. Making this determination requires consideration of several design issues. Further information on many of these issues can be found in epidemiology textbooks, such as those by Kleinbaum and co-workers (1982) and Hennekens and Buring (1987).

Study Types

The primary study designs that are likely to be valuable for addressing the effects of environmental exposures in connection with a dose reconstruction are retrospective cohort studies and case-control studies.

A cohort study is one in which an exposed population is identified and then followed to determine whether specified health effects develop. In a retrospective cohort study, the study is not begun until some period after the initial exposure, but health effects are still identified over the follow-up period of interest, and the data are analyzed in a manner similar to that used in a prospective study. The observed health effects are then compared with health effects expected based on an appropriate control population or related to variations in estimated doses.

In a case-control study, subjects with the disease of interest are identified, and controls are selected, usually matched to cases on the basis of age, gender, and other factors. Estimates of past exposures for cases and controls are then compared. Further discussion of these study designs and their relative advantages and disadvantages is found in Hennekens and Buring (1987) and other epidemiology textbooks.

The choice of study design will depend on several factors, including the health end point of interest, resources for identifying both the exposed population and the cases, whether the exposed population constitutes a large proportion of the population in its geographic location, and the mobility of the population. For exposures that occurred many years ago in mobile populations, it is expected that the retrospective cohort study will usually be preferred over the case-control study. This is because many exposed subjects might have left the area where the exposure occurred, and it is thus likely to be difficult to find a means of identifying cases among these subjects without first identifying the population at risk. Registries, hospitals, and other potential case sources in the area of interest will fail to identify cases who have left the area, and a large proportion of the reported cases will probably be persons who were not in the area at the time of the exposure. However, conducting a case-control study within a retrospective cohort study often is useful for obtaining more detailed information on nonradiation exposures, lifestyle factors, and the like.

Another type of study that might be considered under some circumstances is the correlation or "ecologic" study. In such studies, disease rates are compared for groups of subjects for whom exposures are judged to differ, and the groups are usually defined according to locations for which mortality or morbidity rates are available. Because these studies make use of available statistics, they often can be conducted quickly and inexpensively. Ecologic studies are limited to the use of groups for which disease statistics are available, and this often does not provide the best delineation of exposure, particularly as current geographic location might not reflect past exposure. The use of grouped data can introduce biases especially if the regions chosen reflect social and economic differences; it is often impossible to control for potential confounders in an ecologic study because data are not available or because of limitations in the models that can be used to adjust for grouped data. With data on geographic groups there usually is no assurance that the diseased persons were even exposed. For these reasons, ecologic studies are usually regarded as hypothesis generating, at best, and their results must be regarded as questionable until confirmed with cohort or case-control studies. Problems with the ecologic study design are discussed in more depth by Piantadosi and co-workers (1988), Greenland and Morgenstern (1990), and Greenland (1992).

Although it is not a formal epidemiologic study, interest in possible health effects resulting from a particular exposure source is sometimes generated through identification of a cluster of cases of a specific disease in a particular location, time period, or both. Most clusters are chance occurrences, but it can be difficult to evaluate whether a particular cluster can reasonably be attributed to chance and even more difficult to communicate to the public the role of chance or the magnitude of the purported risks (Slovic 1987). The evaluation of whether a cluster represents an excess involves comparison with a control; often, this comparison is made inappropriately if it is made at all. Also, clusters are sometimes reported without adequate verification of disease status. Nevertheless, clusters require investigation, and if the cluster cannot be readily explained as resulting from chance or from a problem in methodology, a cohort or case-control study might be needed (Kheifets 1993). Problems with cluster studies are discussed in more detail by Rothman (1990) and Neutra (1990).

Statistical Power

The potential informativeness of a study often is measured in terms of statistical power, which can roughly be defined as the probability of rejecting the hypothesis of no effect (null hypothesis) when in fact it is false and the alternative hypothesis is correct. An example would be that one would conclude that there is no increased rate of disease in the exposed group when, in fact, the radiation exposure does have an effect of the expected magnitude. For the studies of interest here, there are two aims that are of primary interest. The first pertains to the probability of detecting a dose-related effect if an effect is present, given the expected size of that effect as derived from the risk estimates that form the basis of radiation protection standards (UNSCEAR 1988, 1993, 1994; NRC 1990; ICRP 1991). To gauge the power of a study for increasing knowledge about radiation effects, it is often meaningful to express the expected detectable effect as a multiple of current radiation risk estimates. For addressing the general concerns of the public, it also is useful to state the effect as a simple excess risk, either relative to baseline risk or as an absolute increase in lifetime risk.

Second, even when a study does not detect an effect, it can yield valuable information if it establishes an upper bound on risks that will increase confidence in the radiation-standard risk estimates or if it can be used to assure the public that the risks have not been large. Thus, a second null hypothesis of interest is that a health effect is not any greater than a specified size (such as the size expected by extrapolation from high-dose data on which the radiation protection standards are based). In practice, this approach usually yields results similar to those obtained by applying the approach described in the previous paragraph, and the two approaches together can be regarded as distinguishing no effect from some specified large effect. If a study has inadequate power to distinguish no effect from effects that would result if risks were several times those based on high-dose extrapolation, the study usually will be judged unlikely to provide a valid and defensible risk estimate. Greenland (1988) provides a discussion of power calculations for distinguishing among various hypotheses, including their relationship to confidence intervals.

To evaluate a study's power, it is necessary to have information on the magnitude of the dose, the size of the population, and the number of cases expected if there were no exposure. The last category requires information on age, sex, length of follow-up, and possibly other factors, and it is determined using available disease rates. At the time the power calculations are made, there is likely to be uncertainty both about the doses and about the size and characteristics of the exposed population. For this reason it is desirable to carry out power calculations based on several alternative assumptions.

The power of a study can depend strongly on the quality of the dose estimates. The lowest power will result if only an average dose can be assigned and if analysis consists of a simple comparison of the exposed group with an appropriate control group judged to be unexposed. Because uncertainties in estimated doses are likely to be large, it is probably desirable to include this scenario as one of those evaluated. However, power often will be much greater if the information gained through the dose reconstruction is incorporated by including several exposure categories and testing for an increase in risk with an increase in dose (Shore and others 1992). Thus, in most cases, power calculations based on the use of the available quantitative information on dose also should be made. Ideally, these calculations would account for errors in the dose estimates (and thus misclassifications of dose); if this is not done, random errors in the measurement of doses will reduce power, can lead to an underestimation of any effect, and can introduce a spurious curvilinearity in the apparent dose-response relationship. Other study limitations, such as the failure to secure participation from all potential subjects and the failure to ascertain all cases, must be considered in developing power calculations. Howe and Chiarelli (1988) describe an approach to power calculations that uses information on dose and allows for the possibility of dose misclassification.

The sample sizes required to achieve adequate statistical power increase greatly as the magnitude of the dose decreases. To illustrate the relationship among dose, sample size, and statistical power, the committee calculated the sample sizes needed to detect an excess in total cancer mortality for various possible doses. The same assumptions were used as in the example earlier in this chapter. It is assumed that each member of a population with an age and sex distribution typical of the United States as a whole was exposed to a specific radiation dose and followed over a full lifetime, and the risk model recommended by BEIR V (NRC 1990) was used to project lifetime risk for radiation-induced fatal cancers (about 7.5 per 10 millisieverts (mSv) per 10,000 persons or 7.5 per rem per 10,000 persons) as compared with about 1,800 spontaneous cancer deaths per 10,000 persons. Table 7-2 shows the size of the exposed population needed at various doses to have an 80% chance of seeing an excess in a comparison against general population rates. It is notable that at doses of 20 mSv (2 rem) or less the required sample sizes are prohibitively large, ranging from 500 thousand to 32 million persons. The sample sizes come into the realm of possibility only when the mean dose is above 50 mSv (5 rem).

Total cancer is not necessarily the most sensitive health end point or the one of most interest for detecting a radiation-induced excess. Table 7-3 shows the required numbers of exposed persons for two cancer types that could be of concern in dose reconstruction studies: leukemia and respiratory cancer mortality. The assumptions are the same as those used above. The BEIR V (NRC 1990) risk models for respiratory cancer and leukemia were used to generate the expected excess malignancies. The results in Table 7-3 show that the required sample sizes are extremely large if the doses are low.

Several caveats should be made regarding these sample size calculations. Normally, one would have a range of doses in the population rather than a uniform dose. Performing a dose-response analysis would create some gain in statistical power (and a corresponding reduction in the required sample size) relative to the simple comparison of the total exposed group to the general population used above (Shore and others 1992). On the other hand, three factors would tend to diminish statistical power. First, the uncertainty in estimating individual doses tends to diminish statistical power and increase the required sample size (Walker and Blettner 1985, De Klerk and others 1989, Armstrong 1990). Second, the calculations in Tables 7-2 and 7-3 assume a full lifetime follow-up.

TABLE 7-2 Size of an Exposed Group Required to Detect an Increase in Total Cancer Mortality with a Lifetime Follow-up, According to Dose

Mean Whole-Body Dose (mSv)	Excess Cancers per 10,000	Sample Size
2.5	1.9	32,000,000
5.0	3.8	7,900,000
10.0	7.5	2,000,000
20.0	15.0	500,000
30.0	22.5	220,000
40.0	30.0	130,000
50.0	37.5	80,000
60.0	45.0	56,000
70.0	52.5	41,000
80.0	60.0	31,000
90.0	67.5	25,000
100.0	75.0	20,000
120.0	90.0	14,000
150.0	113.0	9,100
200.0	150.0	5,200

NOTE: The calculations are based on the BEIR V (NRC 1990, p. 172) estimate of 7.5 excess cancer deaths per 10 mSv per 10,000 persons (averaged across sex and age). For comparison, the number of "spontaneous" background cancer deaths that would be expected in the study population per 10,000 persons is about 1,800. This means, for instance, that at 10 and 100 mSv the radiogenic risks are only about 0.4% and 4%, respectively, as great as the spontaneous cancer mortality. BEIR V used *Vital Statistics of the United States 1980* for the source of baseline data on cancer mortality.

Sample sizes were rounded to two significant digits in view of the approximations involved in their calculation. The table assumes the age and sex-structure and the background cancer rates of the exposed group are comparable to the U.S. general population at the time of irradiation; members of the exposed group are followed up for their remaining lifetimes; the excess cancer risk corresponds to the estimates given by the BEIR V report. The calculations are predicated on achieving 80% statistical power with a 5% alpha level and a one-sided statistical test.

Studies will typically have a much shorter average follow-up than this, which will diminish the statistical power considerably (especially because a large fraction of the cancers would occur at older ages). Third, the publishers of radiation risk assessment studies (NCRP 1980, NRC 1990, ICRP 1991) generally agree that gamma or beta irradiation delivered at low doses and low dose rates probably causes from 2 to 10 times fewer cancer cases per millisievert than do higher, acute doses. Hence, for the non-leukemic cancers included in Tables 7-2 and 7-3, the effects could be overestimated, because they are extrapolated from high-dose studies. For leukemia, however, this low-dose effect is already taken into account by

TABLE 7-3   The Required Size of an Exposed Group to Detect an Increase in Leukemia or Respiratory Cancer Mortality with a Lifetime Follow-up, According to Dose Levels

Mean Target Organ Dose (mSv)	Sample Size	
	Leukemia	Respiratory Cancer
2.5	74,000,000	>100,000,000
5	19,000,000	44,000,000
10	4,700,000	11,000,000
20	1,200,000	2,700,000
30	520,000	1,200,000
40	300,000	680,000
50	190,000	440,000
60	130,000	310,000
70	99,000	220,000
80	76,000	170,000
90	61,000	140,000
100	49,000	110,000
120	25,000	77,000
150	11,000	50,000
200	3,900	28,000

NOTE: The calculations are based on the BEIR V (NRC 1990, p. 175) estimates of excess cancer deaths per 10 mSv per 10,000 persons (averaged across sexes and age) of 0.95 for leukemia and 1.7 for respiratory cancer. It is assumed that the BEIR V estimate of leukemia risk at 100 mSv extrapolates linearly (i.e., there is no quadratic term) downward to lower doses. This has the effect of increasing the estimated low-dose risk and yielding smaller sample sizes than strict adherence to the model would do. See footnote to Table 7-2 for other assumptions.

the linear-quadratic model that was used. Considering all these factors together, the sample sizes in Tables 7-2 and 7-3 probably err on the side of underestimating the required sample sizes.

Outcomes

The health outcomes to be studied should be chosen in keeping with information available about which health effects are expected derived primarily from other studies of exposed populations. As previously noted there is a considerable body of literature on radiation effects, based on animal and human studies. The radionuclides involved in the exposure and the organs of the body that are likely to be most highly exposed are clearly important determinants of the health outcomes chosen for study. With whole-body exposure, cancers of many types could be possible outcomes; in this case, it is important to select a small number, associated with the most radiosensitive sites, as those of primary interest. Public pressure to study effects that have not been strongly linked with radiation in high-dose studies should be resisted because they are likely to be subject to the same biases as cluster studies (Rothman 1990) and to yield spurious results.

Population Identification and Follow-Up

In a cohort study, it is necessary to define the population to be studied carefully to develop the exposure data for this population and to determine whether subjects have developed the health effects of interest. If a population was exposed many years ago, and if a substantial proportion of the population has subsequently moved to other locations, identification of the group will be extremely difficult, if not impossible. Methods for ascertaining health effects will depend on the effect being studied. For diseases with high fatality rates, the use of mortality data from the National Death Index, state death records, and Social Security Administration records could be an option. Tumor registries are another potential source, although with mobile populations and no national registry, this is difficult. For some end points, it will be necessary to locate all members of the population to determine whether health effects have occurred; in some cases (for example, in the study of thyroid disease), it could be necessary to conduct physical examinations of the study population to allow reasonably unbiased evaluation of health effects. Before undertaking an epidemiologic study, it is essential to determine the feasibility of identifying the study population and of ascertaining whether the health effects of interest have occurred. These difficulties could preclude a study regardless of the study's potential statistical power.

Bias and Confounding

Results of an epidemiologic study can be biased for several reasons (bias that results from uncertainties in dose estimates is discussed in Chapter 5 in Uncertainty). For studies conducted in connection with a dose reconstruction, the information needed to estimate dose and the information on confounders often is collected through interviews with subjects or relatives of subjects. The information can be biased both by the magnitude of exposure and by whether the subject has developed the health effect of interest. This could be particularly true if the study has received wide publicity and subjects are aware of the expected health effects and the parameters that affect exposure. It is thus important that questionnaires be designed and administered to minimize bias. Careful attention should be given to the wording of questions and to ensuring, to the extent possible, that interviewers do not know the exposure or disease status of subjects.

Bias also can result if case ascertainment differs by magnitude of exposure, and the study must be designed to minimize such bias. This means that case ascertainment efforts and methods cannot depend on exposure, and it generally rules out the use of volunteer subjects, or information volunteered about disease status. Also, it is important to achieve a high rate of participation in the study; incomplete participation increases the likelihood of bias. When the study involves physical examinations, an effort should be made to prevent the examiners from knowing the exposure status of the examinees.

Even if results are not biased for the methodological reasons noted above, epidemiologic studies are always subject to the possibility of confounding; that is, bias can result from differences among subjects in risk factors other than the exposure of interest. Data on known risk factors for the disease of interest should be collected if possible and taken into account in the analysis. Even when this is done, the possibility of bias cannot be excluded because all possible biasing factors have not been measured. This is an inherent limitation of observational as opposed to experimental studies. (Randomization in experimental studies ensures that the study groups will be comparable, on average.) In the studies considered here, exposed subjects generally reside in specific geographic areas and can differ in various ways (other than exposure) from subjects in other locations. In this regard, comparisons by magnitude of exposure could be less subject to bias than will be comparison with a control population that resides in a different community. However, there might have been socioeconomic gradients by distance from a plant, for example, that could produce bias in the results. A special concern in dose reconstruction studies is exposure to chemicals that correlate with radiation exposures.

Smoking is an example of a particularly important confounder for lung cancer and some other cancers. Because even small differences in smoking habits can have a greater influence on lung cancer risks than does the exposure of interest, it is almost never possible to be certain that one has fully adjusted for smoking even when reasonably detailed smoking histories are available. For this reason, if lung cancer is the health effect of primary interest, it might be necessary to apply stricter criteria in determining whether to conduct a study, and such studies should probably not be conducted if smoking data cannot be obtained. On the other hand, if detailed and reliable smoking data are available, then it should be possible to test if the prevalence of upper respiratory diseases in the cohort under study compared to nonsmokers meets statistical significance.

Bias that results from confounding is especially troublesome in studying small increases in risk, where the magnitude of the bias could exceed the magnitude of the excess risk of interest. Furthermore, bias does not decrease with increased sample size (Boice and Land 1982, Monson 1990). If the expected magnitude of the excess risk resulting from radiation does not exceed the baseline risk by more than a few percent, the results of an epidemiologic study cannot be interpreted unambiguously and could therefore have little value regardless of the size of the population.

Statistical Analysis

Statistical analyses should be designed to make optimal use of available information on doses. If doses have been estimated for each study subject individually, this will usually mean that the analysis will be based on several dose categories (or possibly on ungrouped doses) using methods that are sensitive to an increase in risk with an increase in dose. However, because of uncertainty in dose estimates, it also might be desirable to make overall comparisons of the exposed population with an unexposed control population. Analyses should be adjusted for age, sex, and other potential risk factors for which data are available (Gilbert 1982). Preliminary analyses could determine those risk factors that need to be controlled for in the subsequent analyses. Further discussion of statistical methods for cohort studies is given by Breslow and Day (1987).

Although in the interest of thoroughness it might be desirable to conduct more than one type of analysis, it is important to specify in advance a "primary" approach that is to be emphasized in reporting and interpreting results. Similarly, if more than one health end point is to be studied, it is important to specify the main effects that are to be considered in interpreting study findings; this specification would generally be based on effects seen in other studies of radiation effects.

Results should be presented as estimates of risk, and confidence limits should be used to express uncertainty in the estimates. For the purpose of increasing scientific knowledge about the magnitude of radiation effects, risks need to be expressed per unit of exposure in a form that can be readily compared with estimates from other sources, such as those obtained through extrapolation from high-dose studies. It also is useful to provide estimates of excess risk for subjects in various exposure categories, and possibly for the exposed group as a whole. These risk estimates, which may be particularly helpful for those not familiar with the radiation literature, can be expressed relative to the baseline risk or as increases in absolute risk. In presenting risk estimates and confidence intervals, it is important to emphasize that the conventional confidence intervals do not include uncertainty resulting from bias or uncertainties in dose estimates, or from unidentified confounders that have not been measured or used in the analyses. These issues are discussed by Gilbert (1989).

Uncertainty and Misclassification

Statistical analyses should account for uncertainties in dose estimates (Gilbert 1991a, Clayton 1992). In practice, this is extremely difficult, in part because statistical methods for doing so are complex and often require extensive software development. In addition, it is necessary not only to have information about the uncertainty distributions for doses of individual subjects but also to understand the extent to which these uncertainties are correlated across subjects. Because neither uncertainties nor their correlations are known exactly, it is desirable to conduct dose-response analyses based on several assumptions regarding uncertainties. To accomplish this, it is important that statisticians or epidemiologists work closely with dosimetrists to achieve an understanding of the various uncertainty sources and to help to ensure that uncertainties are expressed in the form needed for use in epidemiologic analyses.

Even though it might not be possible to develop a perfect understanding of the correlation structure, it often is possible to separate uncertainty sources that are highly correlated across subjects from uncertainties that are independent across subjects and that primarily reflect subject variability. These two kinds of uncertainty must be treated differently in epidemiologic analyses, and analyses that incorporate highly correlated uncertainties (such as might result from uncertainties in the magnitude of the source term) are generally much simpler to implement than are analyses that address random uncertainties for individual doses.

Although analyses that consider dose estimation uncertainties are desirable, the primary analysis to be emphasized in presenting results probably should not be adjusted for such uncertainties; the results adjusted for uncertainties would then be presented as an elaboration of the main results. If an uncertainty-adjusted analysis is to be used as the primary analysis, this should be clearly stated in advance, and prior specification of the exact form of the uncertainty adjustment to be used should be included.

SUMMARY AND RECOMMENDATIONS

Radiation dose reconstruction provides detailed quantitative information about individual exposures for epidemiologic study of populations near nuclear facilities. The quality and quantity of the dose information are central to any good epidemiologic study. Because different levels of epidemiologic investigation require different amounts of detail and precision in dose estimates and because both the epidemiologic study and the dose reconstruction will require refinements, dosimetry and epidemiologic screening efforts will be most informative if they are done interactively and in parallel. For example, during the scoping phase of the dose reconstruction, epidemiologists could collect information on the population with potential exposures and provide information about the importance of various sources that contributed to the dose.

Acceptance of an epidemiologic study by the public and the scientific community will hinge on the quality of the science and on the perceived thoroughness of the study. These ends will be achieved only if the following general conditions are met.

First, an epidemiologic study must be justified scientifically, and the decision to undertake it should be based on a careful, preliminary scoping study of the site. The scoping study can define, in outline at least, the design, scope, and methods of the epidemiologic study if one is begun.

Second, if a study is justified, it must begin in a timely manner—delaying the epidemiologic investigation could jeopardize its completeness and hence its credibility. If a study cannot be justified, alternative health surveillance strategies that address public concern should be considered. This will usually be the case if a study does not have reasonable statistical power, either to detect effects that are approximately as large as those predicted from high-dose extrapolation or to set useful upper confidence limits if, in fact, there are no statistically demonstrable effects. For instance, developing a cancer registry would permit continued health monitoring of the long-time residents at the site, so as to detect any unusual cancer incidence. Or a screening program could be a useful surveillance mechanism to provide assurance to the target population that its health concerns are being considered.

Third, if the study is to be credible, there must be close interaction between dosimetrists involved in the dose reconstruction and the epidemiologists who will conduct the study. The epidemiologists must understand the details of the dose reconstruction sufficiently well to collect the appropriate information on dietary and lifestyle factors that could contribute to health effects. It is also necessary for statisticians and epidemiologists to work closely with dosimetrists to define the various sources of uncertainty and to help ensure that uncertainties are expressed in a form that will be useful in epidemiologic analyses.

Fourth, the particular health end points to be targeted in an epidemiologic study will be defined primarily by the organ doses that members of the public received (which may vary appreciably when there are radionuclide exposures) and by the radiosensitivity of various organs and tissues. The primary end points will normally be cancer, although in some circumstances other health end points might be considered (such as adverse reproductive outcomes when pregnant women are exposed).

Fifth, the design of a study and the data needed will depend on the purposes to be served. If an epidemiologic study is conducted, the retrospective cohort or case-control designs are usually the methods of choice. Ecologic (correlational) or cluster studies have large potential for bias and can, therefore, be misleading (Stidley and Samet, 1994). They should be avoided wherever possible. Bias resulting from confounding in epidemiologic studies is especially troublesome in studying small increases in risk, because the magnitude of the bias can exceed the magnitude of the excess risk of interest. Careful attention should be given to ways to control potential biases.

Sixth, the statistical analyses should make optimal use of the available dose information, including uncertainties in the dose estimates. A primary approach to analyzing, reporting, and interpreting results should be specified in advance, to avoid the pitfalls of a posteriori analyses. Similarly, the main health end points to be considered in interpreting the study findings should be specified in advance, to avoid a posteriori "chance" findings; this specification would generally be based on the effects that have been seen in other radiation epidemiology studies.

Finally, as previously noted, studies of exposed populations demand that those populations be involved in their design and implementation. Citizens should be involved in every phase of the scoping and decision-making process and in any studies that are performed. Every effort should be made to communicate scientific issues in a fashion that is understandable to lay audiences. An advisory committee, consisting of citizens and impartial scientists, should be established at the outset and invested with oversight powers. This committee should be maintained throughout the study or surveillance program.

These requirements have led the committee to make the following five recommendations:

29. Dosimetric and epidemiologic scoping studies around the sites of nuclear facilities or accidents should be considered, although the extent of such studies might vary from one site to another based on preliminary evidence about exposures, population sizes, and public concern. These studies should be performed interactively and in parallel, because both are needed to inform a decision about further study of the site or for establishing priorities among sites.

30. Epidemiologic and dosimetric assessments should be closely coordinated. It is important to have epidemiologists involved from the outset of any dose reconstruction activity to ensure that the dosimetric information developed is appropriate for epidemiologic decisions and planning.

31. A full-fledged dose reconstruction and epidemiologic study should be proposed only if the scoping studies show that adverse health effects are likely to be statistically detectable, given the probable dose distribution and size of the exposed population.

32. Studies of health end points for which high-dose studies give no clear evidence of an excess should be avoided, because observing true excesses of these end points is biologically implausible. Such studies tend to waste resources and they are uninformative at best and misleading at worst.

33. A statistical power assessment based on a realistic set of assumptions about the dose distribution, population size, and radiation risk coefficients should be part of the scoping phase.

8

Priority Criteria for
Dose Assessment Studies

RELEASES OF RADIOACTIVE materials have occurred in the past at several places throughout the world. At each location, there is likely to be public demand to determine the effect on health, in the past, present, and future, on the surrounding population in a fair and credible manner. However, given the inevitable limits on resources, a procedure should be considered for setting priorities for studying the effects of releases and for identifying those sites at which past exposures have been particularly significant. The choice of sites for study should be based on an ability to bound the magnitude of the releases and to identify with some confidence, the populations at risk. There should be enough flexibility to permit changes in priority to occur with the disclosure or discovery of additional data. This chapter addresses the process of setting priorities using criteria based on scientific evidence. In practice, however, priorities for specific sites also often are influenced by public concern and political considerations.

An early assessment—a scoping analysis—of the relative significance of a particular site for dose assessment should include both radiologic and epidemiologic scoping studies. A scoping analysis requires the establishment of rules that define when a study should lead to a more detailed investigation and when the detail is sufficient for providing an initial indication of potential damage to public health. The criteria need not be absolute, as they are merely intended to help establish a preliminary ranking—and they can vary for different situations or countries. The approach, however, should be generic, so that it can be extended to other hazardous situations, such as the release of toxic chemicals. This was recommended in a recent NRC report (NRC 1995b) that called for an iterative approach to risk assessment of chemicals starting with relatively inexpensive screening techniques. The decision-making process also must retain some flexibility to remain responsive to other concerns, such as those of affected communities or states.

BASIC CRITERIA

The ranking of sites for dose reconstruction and epidemiologic investigation is based on a comparison among the sites of the urgency for such studies and of the potential magnitude of possible health effects. The committee recognizes that ranking can be based on relatively limited data, using simplified models, and should not anticipate the ultimate assessment. The accuracy of the preliminary estimate is expected to be limited, but it will be sufficient to determine which studies should be done. There are several limitations imposed on a comparative study by the wide differences in the facilities involved, which include research laboratories, production facilities, power plants, or other operations. There are appreciable differences in the size of the sites, the magnitude and the duration of the releases, and size of the potentially exposed population. Another significant factor that might need to be considered is whether there are or were any observed or suspected health effects. Evidence of a potential cluster is difficult to interpret, because most investigations of suspected clusters fail to demonstrate that an excess of disease actually exists. Where a real cluster is identified, the cause can seldom be determined (Kheifets 1993). Nevertheless, reported clusters need to be evaluated and could give leads to further research. Quantitative criteria will be used to inform decisions on when to start a study and when to terminate it. And the decision to conduct a study must admit the possibility of a negative outcome obtained by credible and justifiable scientific procedures. There should be a series of decision gates—selected somewhat arbitrarily, but applied consistently—in three areas: the size and structure of the population; the relative size of the estimated dose compared with doses measured at other facilities; and whether anticipated risks exceed regulatory acceptability or are below the doses generally accepted as being of low consequence by authoritative bodies responsible for the protection of human health. Any site for which the scoping study results in low estimates of the feasibility of both radiologic and epidemiologic studies would automatically be ranked low in priority and would be given the least consideration in the first series of studies.

DECISION CRITERIA

The decision or screens for a scoping study must be applied in a consistent and even-handed manner. First, the plausibility of a study must be determined. To establish a credible scenario involving the off-site population, there must be some evidence of significant releases of radioactive materials or corroboration of off-site effects, for example, from environmental measurements.

Second, a detailed study is feasible only if there is an adequate database. Bounds must be established for minimum requirements for information from plant records, public records, or environmental data. There must be enough information to permit the definition and estimation of a source term and a specific period of release.

Third, the statistical requirements for a meaningful epidemiologic assessment must be met and these requirements should be based on the size of the population, estimated doses, quality and availability of records, and other factors.

Fourth, a radiologic assessment should involve an iterative procedure for making dose estimates that increase in detail, once specific minimum dose criteria are exceeded.

The preliminary scoping study used should eliminate some sites, at least initially, from inclusion in any ranking. It is suggested that one criterion used to assess severity—such as that stated in the Federal Registry 10CFR20 of 0.001 Sv/year (0.1 rem/year) maximum dose to any individual at a nuclear site boundary—be multiplied times 70 years for an added whole-body lifetime dose of 0.07 Sv (7 rem) and be used as a realistic selection gate for this purpose, especially if the number of competing sites is high. Alternatively, because internal radionuclides could be the primary source of exposure, the severity criterion could be stated in terms of the absorbed organ dose to those organs at greatest risk.

FINAL RANKING

The final ranking is a relative assessment and is subject to iterative review. Sites that are given a high priority by both the epidemiologic and the radiologic preliminary evaluations would be assigned a high priority for a more detailed dose reconstruction study and epidemiologic feasibility studies. To increase the usefulness and cost-effectiveness of these studies, they should be conducted in parallel and with clear procedures for investigators to follow. Such continuing involvement and interaction ensure that both studies are designed with a common goal and this approach is more likely to ensure that dose reconstruction studies account for such issues as avoidance of bias and evaluating all sources of potential uncertainty. Similarly it ensures that an epidemiologic study will focus on the population most likely to have been affected and will collect all the necessary information for detailed dose reconstruction.

Sites that receive a high rating in the dose assessment but a low one on epidemiologic grounds (or vice versa) would receive a lower priority ranking and would need to be evaluated case by case. Sites that pass the decision criteria, but still rate low on both scales would receive the lowest priority and could be dropped from further detailed studies. For example, for a large population that is concerned but has received relatively low doses, an epidemiologic study designed to detect health effects might not be feasible. In this case, extensive dose reconstruction might not be necessary or justifiable. Instead, the concerns of this population might best be addressed by establishing and maintaining surveillance registries for continuous public health monitoring so that any health effects from any future exposures would be addressed quickly. In the event of high exposures to a small population, comprehensive epidemiologic studies will usually be warranted, but before they begin, additional dose reconstruction could be necessary to provide the best possible estimates of the doses involved and the nature of the exposure (ambient or through ingestion or inhalation of radionuclides).

Finally, it is crucial to maintain a distinction between scientific criteria and other considerations in risk assessments. These other considerations include public concern and the fact that the public might have difficulty in understanding the concepts of statistical significance, uncertainty, and dose-response relationships.

SUMMARY AND RECOMMENDATIONS

Given the inevitable limits on available resources and to avoid capricious judgments in the selection of sites to be studied, a procedure must be devised for setting priorities for dose reconstruction studies and for identifying those sites where past exposures may have been particularly significant. A series of objective criteria should be applied, including the ability to bound the magnitude of the historic releases with some confidence, and the criteria should be sufficiently flexible to permit changes in priority to occur with the disclosure or discovery of additional data. The committee has, therefore, set forth a tentative process of setting priorities using criteria based on scientific evidence. In practice, however, it is recognized that priorities for specific sites also are influenced by other considerations. The committee makes the following three recommendations:

34. Rules and criteria should be defined for a scoping analysis to determine the desirability of a dose assessment and epidemiologic study. These should include demonstration of the feasibility and plausibility of the study, evidence of an adequate database, demonstration of an adequate range of doses, and appropriate numbers of subjects at the higher end of the dose range to meet the statistical needs of an epidemiologic study.

35. Quantitative criteria should be used to arrive at a credible and cost-effective ranking of sites for study.

36. An iterative procedure should be used for making dose estimates that increase in detail after specific minimum dose criteria are exceeded.

9

Conclusions

T HE RECONSTRUCTION IN space and time of doses off-site from the release of radioactive material from DOE-managed nuclear facilities involves several steps (refer to Figure 9-1):

1. An *analysis of the source term* is used to estimate the magnitude of the releases of radionuclides and the periods over which they were released, including episodic releases from nonroutine events.

2. An *analysis of the environmental pathways* examines the transport of the released radionuclides to identify the concentrations in environmental media such as air, water, and food.

3. An *assessment of radiation doses and risks* brings together all of the data on releases, transport, and biologic factors to determine doses to persons and the resulting likelihood of disease in those who have been exposed.

4. An examination of *epidemiologic considerations* is used to evaluate the feasibility and scientific merit of an epidemiologic study.

5. An *uncertainty and sensitivity analysis* of the parameters and values is used to establish confidence intervals and to identify important factors in the overall analysis of the dose reconstruction.

The way the steps are implemented will vary from site to site, but to achieve maximal scientific rigor and some consistency across sites the committee makes the following recommendations:

1. In dose reconstruction studies, thorough consideration should

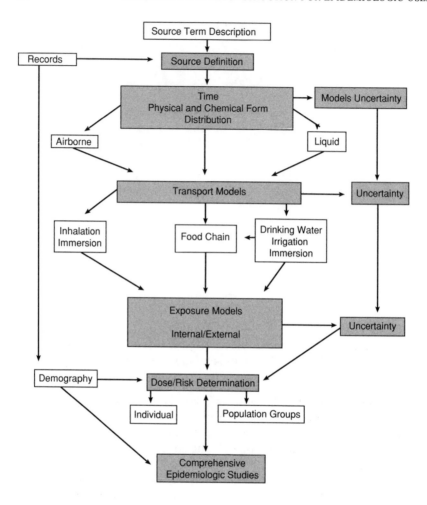

FIGURE 9-1   Dose reconstruction methodology.

be given to the collection of representative data, to an evaluation of their representativeness, to quality control, and to public involvement from the time the decision-making process begins, through the release of any results, and after the study concludes in any follow-up activities.

2. An advisory or steering committee should be established at the outset of a dose reconstruction study. This committee should consist of members of the public and knowledgeable scientists who are not associated with the investigators or their sponsors. The meetings of this committee should be open to the public. Along with its steering duties,

this committee should be charged with responsibility for establishing an interactive process to communicate the elements and conclusions of the study to the general public.

3. Dose reconstruction studies should begin with a scoping study—a preliminary analysis—to determine whether a comprehensive dose reconstruction study is needed or even possible, either for its own purposes or as the basis for a comprehensive epidemiologic study.

4. All dose reconstruction studies should be reviewed by groups of scientists and public health officials who are not directly involved in the study, either as participants or as advisors, and time and resources should be allocated for resolving discrepancies in the results.

5. There should be coordination between the dosimetric and epidemiologic efforts, which should begin at the outset of the dosimetric study and continue throughout.

6. Premature dissemination of the results should be avoided. Results should not be disclosed until the dose assessment study is complete, has undergone peer review, and has been published. Dissemination of data being considered during the study is appropriate and desirable.

7. A clear understanding of the public's concern should be gained before the study begins. A dose reconstruction study should not proceed until the design is such that it is likely that the results will address the public's concern.

8. Scoping studies should be the primary approach to initiating a source term evaluation. These should be followed, if appropriate, by more comprehensive studies. The scoping study of the source term should seek to generate the data needed to identify the environmental pathways of potential importance and to permit estimation of the concentrations of radionuclides to which the public might have been exposed.

9. To ensure maximum confidence in the source term analysis, proprietary or classified information should be made available for the analysis, or a mechanism should be developed to determine whether such data are essential to the accuracy and consistency of the analysis.

10. The source term should be derived chiefly from available original data in as many different ways as practical. The source term should be confirmed, wherever possible, by comparison with independent environmental monitoring data from another source of information.

11. To ensure completeness and accuracy of the estimated source term, all relevant data should be evaluated. Any gaps in the data should be analyzed carefully for their significance and filled by reconstruction from existing data if appropriate.

12. Episodic events should be documented as separate releases for

specific consideration in environmental transport and dose calculations. An event is considered episodic if it lasted for less than 10 days and if the release rate was at least 10 times the average monthly or annual rate.

13. The release quantities provided for use in a comprehensive study of the source term should be complete, unbiased estimates of all amounts and forms of relevant materials released to the environment.

14. Insofar as possible, measurements of environmental radiation or of radionuclides should be used in the environmental pathway analysis. For example, if a single contaminating event has taken place and if measurements have been made (such as external gamma exposure rate or deposition of one or more radionuclides or stable materials), it is often possible to begin the dose reconstruction without the need to model the transport of radionuclides up through this stage. Even if the contamination is chronic, it is often preferable to take suitable soil samples to measure the longer lived components of the contaminant materials (such as ^{137}Cs or ^{129}I) and to infer the deposition of shorter lived components (such as ^{131}I) rather than to depend on atmospheric transport and deposition models, which are much less reliable than are direct measurements.

15. Even if there is an abundant base of monitoring data, mathematical models are usually needed to extrapolate information from situations where measurements have been made to situations where measurements are lacking. Every attempt should be made to validate the predictions of the models against relevant data sets. Caution should be exercised in the use of "off-the-shelf" computer codes that may have been developed for other purposes such as regulatory analyses.

16. Environmental pathway analyses should include quantitative estimates of uncertainty to indicate the degree of confidence that can be placed in exposure estimates.

17. For accidents, there should be careful scrutiny of any countermeasures, such as removal of contaminated foodstuffs from commerce. Even if contaminated foodstuffs are not removed, people often voluntarily avoid contaminated foodstuffs and change their dietary habits. For routine releases, attention should be paid in the assessment of ingested radionuclides to the movement of foodstuffs into the region of interest because people do not necessarily consume local agricultural products.

18. All exposures from external sources, inhaled radionuclides, and ingested radionuclides should be considered; when certain pathways or other factors suggest that a particular source term or radionuclide will not contribute substantial dose, reports should explain why these sources or specific radionuclides were not considered in the final estimations.

19. Dose assessment should proceed at three levels: preliminary, comprehensive, and individual dose assessment.

20. Acceptable levels of uncertainty should be defined before a decision is made to carry out a detailed study.

21. A readily available set of intake-to-annual dose conversion factors for long-lived radionuclides should be established.

22. Doses should be expressed as effective doses in a preliminary dose assessment and objective criteria should be used to decide whether it is warranted to embark on a full-fledged study.

23. Dose estimates in a comprehensive dose assessment should be expressed as the annual organ absorbed doses from low-LET and high-LET radiations. Estimation of the effective dose may also be helpful.

24. In an individual dose assessment, the doses should be described separately as the annual organ-absorbed doses from low-LET and high-LET radiations.

25. For biologic markers to be useful in dose reconstruction, research will be necessary to

- measure the stability of persistent biologic markers,
- define "calibration curves" for low to moderate and chronic exposures,
- determine the frequency of specific markers in unexposed populations,
- define the sources of inter-individual variability for various markers, and
- develop better definitions of marker responses after partial-body (external) or specific organ (internal emitter) exposure.

26. New assays should be developed to address the problems with individual variability in background, with identification of differences in individual susceptibility to radiation genotoxicity, and with the lack of sensitivity for quantifying low radiation exposures so that acute doses greater than 0.1 Gy (10 rad) can be reconstructed.

27. Biologic markers of effect should be used as epidemiologic end points. However, until clear connections are established between the marker and the disease, their use could be misleading rather than illuminating.

28. As the utility of biologic markers becomes established and accepted, the committee recommends that the CDC develop procedural strategies for conducting field studies for both specimen collection and laboratory analyses in the event of an acute release of activity.

29. Dosimetric and epidemiologic scoping studies around the sites of nuclear facilities or accidents should be considered, although the extent of such studies might vary from one site to another based on preliminary evidence about exposures, population sizes, and public

concern. These studies should be performed interactively and in parallel, because both are needed to inform a decision about further study of the site or for establishing priorities among sites.

30. Epidemiologic and dosimetric assessments should be closely coordinated. It is important to have epidemiologists involved from the outset of any dose reconstruction activity to ensure that the dosimetric information developed is appropriate for epidemiologic decisions and planning.

31. A full-fledged dose reconstruction and epidemiologic study should be proposed only if the scoping studies show that adverse health effects are likely to be statistically detectable, given the probable dose distribution and size of the exposed population.

32. Studies of health end points for which high-dose studies give no clear evidence of an excess should be avoided, because observing true excesses of these end points is biologically implausible. Such studies tend to waste resources and they are uninformative at best and misleading at worst.

33. A statistical power assessment based on a realistic set of assumptions about the dose distribution, population size, and radiation risk coefficients should be part of the scoping phase.

34. Rules and criteria should be defined for a scoping analysis to determine the desirability of a dose assessment and epidemiologic study. These should include demonstration of the feasibility and plausibility of the study, evidence of an adequate database, demonstration of an adequate range of doses, and appropriate numbers of subjects at the higher end of the dose range to meet the statistical needs of an epidemiologic study.

35. Quantitative criteria should be used to arrive at a credible and cost-effective ranking of sites for study.

36. An iterative procedure should be used for making dose estimates that increase in detail after specific minimum dose criteria are exceeded.

10

Literature Cited

Akiyama M, Kusunoki Y, Umeki S, Hirai Y, Nakamura N, Kyoizumi S. 1992. Evaluation of four somatic mutation assays as biological dosimeter in humans. In: Dewey WC, Edington M, Fry RJM, Hall EJ, Whitmore GF, editors. Radiation research: A twentieth-century perspective. Volume II: Congress proceedings. San Diego, California: Academic Press. p 177-182.

Albertini RJ, Castle KL, Borcherding WR. 1982. T-cell cloning to detect the mutant 6-thioguanine-resistant lymphocytes present in human peripheral blood. Proc Natl Acad Sci (USA) 79:6617-6621.

Albertini RJ, Nicklas JA, O'Neill JP, Robison SH. 1990. In vivo somatic mutations in humans: measurement and analysis. Annual Review of Genetics 24:305-326.

Anspaugh LR, Church BW. 1986. Historical estimates of external exposure and collective external γ exposure from testing at the Nevada Test Site. I. Test series through Hardtack II 1958. Health Phys 51:35-51.

Anspaugh LR, Ricker YE, Black SC, Grossman RF, Wheeler DW, Church BW, Quinn VE. 1990. Historical estimates of external γ exposure and collective external γ exposure from testing at the Nevada Test Site. II. Test series after Hardtack II, and summary. Health Phys 59:525-532.

Armstrong B. 1990. Effects of measurement errors on estimates of exposure-response relationships. Recent Results Cancer Res 120:50-63.

Beck HL. 1980. Exposure rate conversion factors for radionuclides deposited on the ground. U.S. Department of Energy, Environmental Measurements Laboratory, Report EML-378.

Beck HL, Anspaugh LR . 1990. Development of the county data base. Estimates of exposure rates and times of arrival of fallout in the ORERP phase-II area. Comparison with cumulative deposition estimates based on analyses of retrospective and historical soil samples. U.S. Department of Energy, Las Vegas Operations Office, Report NVO-320.

Beck HL, Krey PW. 1983. Radiation exposures in Utah from Nevada nuclear tests. Science 220:18-24.

Bender M, Gooch PC. 1967. Chromosome aberrations in irradiated humans. In: Turano L, Ratti A, Biagini C, editors. Progress in radiobiology, Volume II. Amsterdam: Excerpta Medica Foundation. p 1421-1425.

Boice JD, Land C. 1982. Ionizing radiation. In: Schottenfeld D, Fraumeni J, editors. Cancer epidemiology and prevention. Philadelphia, Pa: WB Saunders Co. p 231-153.

Bouville A, Dreicer M, Beck HL, Hoecker WH, Church BW. 1990. Models of radioiodine transport to populations within the continental U.S. Health Phys 59:659-668.

Brandão-Mello CE, Oliverira AR, Valverde NJ, Farina R, Cordeiro JM. 1991. Clinical and hematological aspects of ^{137}Cs: The Goiânia radiation accident. Health Phys 60:31-39.

Brenk HD, Fairobent E, Markee EH, Jr. 1983. Transport of radionuclides in the atmosphere. In: Till JE, Meyer HR, editors. Radiological assessment: a textbook on environmental dose analysis. U.S. Nuclear Regulatory Commission, Report NUREG/CR-3332 p 2-1–2-86.

Breslow NE, Day NE, editors. 1987. Statistical methods in cancer research. Volume II: The design and analysis of cohort studies. New York, NY: Oxford University Press.

Bureau International des Poids et Mesures. 1991. Le systém international d'unités (SI) 6th ed. Bureau International des Poids et Mesures.

Cate S, Ruttenber AJ, Conklin AW. 1990. Feasibility of an epidemiologic study of thyroid neoplasia in persons exposed to radionuclides from the Hanford nuclear facility between 1944 and 1956. Health Phys 59:169-178.

Cederwall RT, Ricker YE, Cederwall PL, Homan DN, Anspaugh LR. 1990. Ground-based air-sampling measurements near the Nevada Test Site after atmospheric nuclear tests. Health Phys 59:533-540.

Church BW, Wheeler DL, Campbell CM, Nutley RV, Anspaugh LR. 1990. Overview of the Department of Energy's Off-site Radiation Exposure Review Project (ORERP). Health Phys 59:503-510.

Clayton DG. 1992. Models for the analysis of cohort and case-control studies with inaccurately measured exposures. In: Dwyer JH, Lippert P, Feinleib M, Hoffmeiser H, editors. Statistical models for longitudinal studies of health. New York, NY: Oxford University Press. p 301-331.

De Klerk NH, English D, Armstrong B. 1989. A review of the effects of random measurement error on relative risk estimates in epidemiological studies. Int J Epidemiol 18: 705-712.

Dolphin GW, Lloyd DC, Purrott RJ. 1973. Chromosome aberration analysis as a dosimetric technique in radiological protection. Health Phys 25:7-15.

Eckerman KE, Ryman JC. 1993. External exposure to radionuclides in air, water, and soil. Federal Guidance Report 12, EPA 402-R-93-081. Washington, DC: Environmental Protection Agency.

Engelmann RJ. 1970. Scavenging prediction using ratios of concentration in air and precipitation in precipitation scavenging. CONF-700601.

Evans HJ, Buckton KE, Hamilton GE, Carothers A. 1979. Radiation-induced chromosome aberrations in nuclear-dockyard workers. Nature 277:531-534.

Gavrilyan I, Gordeyev KI, Ivanov UK, Ilyin LA, Kondrusev AI, Margulis U, Stepanenko VF, Khrushch VT, Shikarev SM. 1992. Estimation of thyroid radiation according to the results of radiation internal monitoring of the Byelorussian population after the Chernobyl accident: Features and results. Vestn Akad Med Nauk SSSR. 2:35-43.

Gilbert ES. 1982. Some confounding factors in the study of mortality and occupational exposures. Am J Epidemiol 116:177-188.

Gilbert ES. 1989. Issues in analysing the effects of occupational exposure to low levels of radiation. Stat Med 8:173-187.

Gilbert ES. 1991a. Accounting for bias and uncertainty resulting from dose measurement errors and other factors. Br Inst Radiol Rpt 22:155-159.

Gilbert ES. 1991b. Late somatic effects. In: Abrahamson S, Bender MA, Boecker BB, Gilbert ES, Scott BR, editors. Health effects models for nuclear power plant accident consequence analysis. Modification of models resulting from recent reports on health effects of ionizing radiation. Low LET radiation. Part II. Scientific basis for health effects models. NUREG/CR-4214. Rev. 1, Part II, Addendum 1, LMF-132.

Gilbert ES, Cragle D, Wiggs L. 1993. Updated analyses of combined mortality data on workers at the Hanford site, Oak Ridge National Laboratory, and Rocky Flats weapons plant. Radiat Res 136:408-421.

Godoy JM, Guimaraes JRD, Pereira JCA, Pires do Rio MA. 1991. Cesium-137 in the Goiania waterways during and after the radiological accident. Health Phys 60:99-103.

Gooch PC, Bender MA, Randolph ML. 1964. Chromosome aberrations induced in human somatic cells by neutrons. In: Biological effects of neutron and proton irradiations, Volume I. Vienna: International Atomic Energy Agency. p. 325-342.

Grant SG, Bigbee WL. 1993. In vivo somatic mutation and segregation at the human glycophorin A (GPA) locus: Phenotypic variation encompassing both gene-specific and chromosomal mechanisms. Mutat Res 288:163-172.

Grant SG, Jensen RH. 1993. Use of hematopoietic cells and markers for detection and quantitation of human in vivo somatic mutation. In: Garratty G, editor. Immunobiology of transfusion medicine. New York, NY: Marcel Dekker, p. 299-323.

Gray JW, Kallioniemi A, Pinkel D, Tkachuk D, Weier HU, Lucas J, Kallioniemi O, Straume T, Tenjin T, Kuo WL. 1991. Applications of fluorescence in situ hybridization. In: Gledhill BL, Mauro F, editors. Biological dosimetry and detection of disease-specific chromosome aberrations. New York: Wiley-Liss. p. 399-401.

Greenland S. 1987. Quantitative methods in the review of epidemiologic literature. Epidemiol Rev 9:1-30.

Greenland S. 1988. On sample-size and power calculations for studies using confidence intervals. Am J Epidemiol 128:231-237.

Greenland S. 1992. Divergent biases in ecologic and individual-level studies. Stat Med 11:1209-1223.

Greenland S, Morgenstern H. 1990. Ecological bias, confounding, and effect modification. Int J Epidemiol 19:764-767.

Grossman RF, Thompson CB. 1990. The survey meter database: Results of radiation monitoring near the Nevada Test Site after nuclear tests. U.S. Department of Energy, Las Vegas Operations Office.

Hammond SE. 1971. Industrial-type operations as a source of environmental plutonium. In: Fowler EB, Henderson RW, Milligan MR, editors. Proceedings of Environmental Plutonium Symposium, Los Alamos, New Mexico: Los Alamos National Laboratory, LA-4756. p 25-35.

Hanna SR, Briggs GA, Hosker RP, Jr. 1982. Handbook on atmospheric diffusion. Technical Information Center, U.S. Department of Energy.

Hardy EP, Krey PW. 1971. Determining the accumulated deposit of radionuclides by soil sampling and analysis. In: Fowler EB, Henderson RW, Milligan MR, editors. Proceedings of Environmental Plutonium Symposium. Los Alamos, New Mexico: Los Alamos National Laboratory, LA-4756. p 37-42.

Haskell EH, Bailiff IK, Kenner GH, Kaipa PL, Wrenn ME. 1994. Thermoluminescence measurements of gamma-ray doses attributable to fallout from the Nevada Test Site using building bricks as natural dosimeters. Health Phys 66:380-391.

Health Physics. 1991. The Goiania radiation accident. Health Phys (Special issue) 60:1-113.

Heeb CM. 1992. Iodine-131 releases from the Hanford site, 1944 through 1947, Report PNWD-2033. Richland, Washington: Battelle, Pacific Northwest Laboratories.

Heeb CM, Morgan LG. 1991. Iodine-131 in irradiated fuel at time of processing from December 1944 through December 1947, PNL-7253 HEDR, Richland, Washington: Battelle, Pacific Northwest Laboratories.

Henderson RW, Smale RF. 1990. External exposure estimates for individuals near the Nevada Test Site. Health Phys 59:715-721.

Hennekens CH, Buring JE. 1987. Epidemiology in medicine. Boston: Little and Brown.

Hicks HG. 1982. Calculation of the concentration of any radionuclide deposited on the ground by off-site fallout from a nuclear detonation. Health Phys 42:585-600.

Hicks HG. 1990. Additional calculations of radionuclide production following nuclear explosions and Pu isotopic ratios for Nevada test events. Health Phys 59:515-523.

Hoffman FO. 1993. Analysis of exposure pathways for dose reconstruction and epidemiology. Presented at the National Academy of Sciences Workshop on Dose Reconstruction for Epidemiological Uses. October 25-27, 1993. Washington, DC.

Howe GR, Chiarelli AM. 1988. Methodological issues in cohort studies. II: Power calculations. Int J Epidemiol 17:464-468.

[IAEA] International Atomic Energy Agency. 1989. Safety practice publication of the IAEA: Evaluating the reliability of predictions using environment transfer models. Safety Series Number 100. STI/PUB/835. Vienna: International Atomic Energy Agency. p. 1-106.

[ICRP] International Commission on Radiological Protection. 1979. Limits for intakes of radionuclides by workers. International Commission on Radiological Protection Publication 30 Part 1. Annals of the ICRP. 2(3/4). Pergamon Press, Oxford, England.

[ICRP] International Commission on Radiological Protection. 1989. Age-dependent doses to members of the public from intake of radionuclides. ICRP Publication 56, Oxford: Pergamon Press.

[ICRP] International Commission on Radiological Protection. 1991. 1990. Recommendations of the International Commission on Radiological Protection. Ann Intl Comm Radiol Protect 21:1-201.

[ICRP] International Commission on Radiological Protection. 1995. Human respiratory tract model for radiological protection. A report of Committee 2 of the ICRP. Oxford: Pergamon.

Janatipour M, Trainor KJ, Kutlaca R, Bennett G, Hay J, Turner DR, Morley AA. 1988. Mutations in human lymphocytes studied by an HLA selection system. Mutat Res 198:221-226.

Jirka GH, Findikakis AN, Onishi Y, Ryan PJ. 1983. Transport of radionuclides in surface waters. In: Till JE, Meyer HR, editors. Radiological assessment. A textbook on environmental dose analysis. U.S. Nuclear Regulatory Commission, Report NUREG/CR-3332.

Kendall GM, Muirhead CR, MacGibbon BH, O'Hagan JA, Conquest AJ, Goodill AA, Butland BK, Fell TP, Jackson DA, Webb MA, Haylock RGE, Thomas JM, Silk TJ. 1992. Mortality and occupational exposure to radiation: First analysis of the national registry for radiation workers. Br Med J 304:220-225.

Kerber RA, Till JE, Simon SL, Lyon JL, Thomas DC, Preston-Martin S, Rallison ML, Lloyd RD, Stevens W. 1993. A cohort study of thyroid disease in relation to fallout from nuclear weapons testing. JAMA 270:2076-2082.

Kheifets L. 1993. Cluster analysis: a perspective. Statistics in Medicine 12:1755-1757.

Kleinbaum DG, Kupper L, Morgenstern H. 1982. Epidemiologic research: Principles and quantitative methods. Belmont, CA: Wadsworth.

Kossenko MM. 1991. Medical effects of population exposure to radiation as a result of radiation accidents in the Southern Urals. Abstract. Dissertation for the Degree of Medical Sciences. Institute of Biophysics, USSR Ministry of Health. Moscow.

Kossenko MM, Degteva M. 1994. The follow-up of the population exposed as a result of the release of radioactive wastes into the Techa River. Part 3. Cancer mortality and risk evaluation. Sci Tot Environ.

Kossenko MM, Degteva MO, Petrushova NA. 1992. Estimate of the risk of leukemia to residents exposed to radiation as a result of a nuclear accident in the Southern Urals. The PSR Quarterly 2:187-197.

Kyoizumi S, Nakamura N, Hakoda M, Awa AA, Bean MA, Jensen RH, Akiyama M. 1989. Detection of somatic mutations at the glycophorin A locus in erythrocytes of atomic-bomb survivors using a single beam flow sorter. Cancer Res 49:581-588.

Land CE, Sinclair WK. 1991. The relative contributions of different organ sites to the total cancer mortality associated with low-dose radiation exposure. In: Risks associated with ionising radiation. Annals of the ICRP 22, No. 1.

Langer G. 1989. Resuspension of soil particles from Rocky Flats containing plutonium particulates: a review. Rockwell International, Rocky Flats Plant, Box 464, Golden, Colorado.

Langlois RG, Bigbee WL, Kyoizumi S, Nakamura N, Bean MA, Akiyama M, Jensen RH. 1987. Evidence for increased somatic cell mutations at the glycophorin A locus in atomic bomb survivors. Science 236:445-448.

Langlois RG, Akiyama M, Kusunoki Y, DuPont BR, Moore DH, WL Bigbee, Grant SG, Jensen RH. 1993. Analysis of somatic cell mutations at the glycophorin A locus in atomic bomb survivors: a comparative study of assay methods. Radiation Res 136:111-117.

Likhtarev IA, Shandala NK, Gulko GM, Kairo A, Chepurny NI. 1993. Ukrainian thyroid doses after the Chernobyl accident. Health Phys 64:594-599.

Littlefield LG, Sayer AM, Frome EL. 1989. Comparisons of dose-response parameters for radiation-induced acentric fragments and micronuclei observed in cytokinesis-arrested lymphocytes. Mutagenesis 4: 265-270.

Littlefield LG, Joiner EE, Hubner KF. 1990. Cytogenetic techniques in biological dosimetry: Overview and example of dose estimation in 10 persons exposed to gamma radiation in the 1984 Mexican ^{60}Co-accident. In: Mettler FA Jr., Kelsey CA, Ricks RC, editors. Medical management of radiation accidents. Boca Raton, Florida: CRC Press. p 109-126.

Lloyd DC, Purrott RJ, Reeder EJ. 1980. The incidence of unstable chromosome aberrations in peripheral blood lymphocytes from unirradiated and occupationally exposed people. Mutation Res 72:523-532.

Lloyd DC, Edwards AA, Prosser JJ, Barjaktarovic N, Brown JK, Horvot D, Ismail SR, Koteles GJ, Almassy Z, Krepinski A, Kucerova M, Littlefield LG, Mukherjee U, Natarajan AT, Sasaki MS. 1987. A collaborative exercise on cytogenetic dosimetry for simulated whole and partial body accidental irradiation. Mutat Res 179:197-208.

Loevinger R, Budinger TF, Watson EE. 1991. MIRD primer for absorbed dose calculations. New York: The Society of Nuclear Medicine.

Lucas JN, Tenjin T, Straume T, Pinkel D, Moore D, Litt M, Gray JW. 1989a. Rapid determination of human chromosome translocation frequency using a pair of chromosome-specific DNA probes. Int J Radiat Biol 56:35-44.

Lucas JN, Tenjin T, Straume T, Pinkel D, Moore D, Litt M, Gray JW. 1989b. Letter to the Editor. Int J Radiat Biol 56:201.

Lucas JN, Awa A, Straume T, Poggensee M, Kodama Y, Nakano M, Ohtaki K, Weier W, Pinkel D, Gray J, Littlefield G. 1992a. Rapid translocation frequency analysis in humans decades after exposure to ionizing radiation. Int J Radiat Biol 62:53-63.

Lucas JN, Poggensee M, Straume T. 1992b. The persistence of chromosome translocations in a radiation worker accidentally exposed to tritium. Cytogenetics Cell Genetics 60: 255-256.

Lucas JN, Cox AB, McLean J. 1993. Human chromosome "painting" probes used to measure chromosome translocations in non-human primates: extrapolations from monkey to man. Radiation Protection Dosimetry 46:253-256.

Lucas, JN, Swansbury G, Clutterbuck R, Hill F, Burk C, Straume T. 1994. Discrimination between leukemia- and non-leukemia-related chromosomal abnormalities in the patient's lymphocytes. Int J Radiat Biol 66:185-189.

Mabuchi K, Soda M, Ron E, Tokunaga M, Ochikubo S, Sugimoto S, Ikeda T, Terasaki M, Preston DL, Thompson DE. 1994. Cancer incidence in atomic bomb survivors. Part I: Use of the tumor registries in Hiroshima and Nagasaki for incidence studies. Radiat Res 137:S-1-S-16.

Megaw WJ. 1965. The absorption of iodine on atmospheric particles. J Nucl Energy 19:585-595.

Mendelsohn ML, Peeters JP, Normandy MJ, editors. 1995. Biomarkers and occupational health: Progress and perspectives. Washington, DC: Joseph Henry Press.

Monson RR. 1990. Occupational epidemiology. Boca Raton, Florida: CRC Press.

Morley AA, Trainor KJ, Seshadri R, Ryall RG. 1983. Measurement of in vivo mutations in human lymphocytes. Nature 302:155-156.

Nakano M, 1993. Frequency of reciprocal translocations and dicentrics induced in human blood lymphocytes by x-irradiation as determined by fluorescence in situ hybridization. Int J Radiat Biol 64:565-569.

Napier BA. 1992. Determination of radionuclides and pathways contributing to cumulative dose. BN-SA-3673 HEDR. Richland, Washington: Battelle, Pacific Northwest Laboratories.

[NCRP] National Council on Radiation Protection and Measurements. 1980. Influence of dose and its distribution in time on dose-response relationships for low-LET radiations. Washington, DC. National Council on Radiation Protection and Measurements, Report No. 64.

[NCRP] National Council on Radiation Protection and Measurements. 1985. SI units in radiation protection and measurements. Report No. 82. Bethesda, Maryland: National Council on Radiation Protection and Measurements.

[NCRP] National Council on Radiation Protection and Measurements. 1993. Risk Estimates for Radiation Protection. Report No. 115. Bethesda, Maryland: National Council on Radiation Protection and Measurements.

NRC] National Research Council, Committee on Dose Assignment and Reconstruction for Service Personnel at Nuclear Weapons Tests. 1985. Review of the methods used to assign radiation doses to service personnel at nuclear weapons tests. Washington, DC: National Academy Press.

[NRC] National Research Council, Committee on the Biological Effects of Ionizing Radiations. 1990. Health effects of exposure to low levels of ionizing radiation (BEIR V). Washington, DC: National Academy Press.

[NRC] National Research Council, Committee on an Assessment of CDC Radiation Studies. 1992. Dose reconstruction for the Fernald nuclear facility. Washington, DC: National Academy Press.

[NRC] National Research Council, Committee on an Assessment of CDC Radiation Studies. 1994a. Dose reconstruction for the Fernald nuclear facility: A review of task 4. Washington, DC: National Academy Press.

[NRC] National Research Council, Committee on an Assessment of CDC Radiation Studies. 1994b. The Hanford environmental dose reconstruction project: a review of four documents. Washington, DC: National Academy Press.

[NRC] National Research Council, Committee to Review Risk Management in the DoE's Environmental Remediation Program. 1994c. Building consensus through risk management of the Department of Energy's Environmental Remediation Program. Washington, DC: National Academy Press.

[NRC] National Research Council, Committee on an Assessment of CDC Radiation Studies. 1995a. A review of two HEDR dosimetry reports: Columbia River pathway and atmospheric pathway. Washington, DC: National Academy Press.

[NRC] National Research Council, Committee on Risk Assessment of Hazardous Air Pollutants. 1995b. Science and judgment in risk assessment. Washington, DC: National Academy Press.

Neutra RR. 1990. Counterpoint from a cluster buster. Am J Epidemiol 132:1-8.

Ng YC, Anspaugh LR, Cederwall RT. 1990. ORERP internal dose estimates for individuals. Health Phys 59:693-713.

Nicklas JA, O'Neill JP, Albertini RJ. 1986. Use of T-cell receptor gene probes to quantify the in vivo hprt mutations in human T-lymphocytes. Mutation Res 173:67-72.

Nikipelov B, Lyzlov A, Koshurnikova N. 1990. An experience of the first enterprise of the nuclear industry (levels of exposure and health of workers). Priroda 2:30-38.

Petoussi N, Jacob P, Zankl M, Saito K. 1991. Organ doses for foetuses, babies, children, and adults from environmental gamma rays. Radiat Protection Dosimetry 37:31-39.

Piantadosi S, Byar DP, Green SB. 1988. The ecological fallacy. Am J Epidemiol 127:893-904.

Pinkel D, Landegent J, Collins C, Fuscoe J, Segraves R, Lucas JN, Gray J. 1988. Fluorescence in situ hybridization with human chromosome-specific libraries: Detection of trisomy 21 and translocations of chromosome 4. Proc Natl Acad Sci (USA) 85:9138-9142.

Preston DL, Kusumi S, Tomonaga M, Izumi S, Ron E, Kuramoto A, Kamada N, Dohy H, Matsui T, Nonaka H, Thompson DE, Soda M, Mabuchi K. 1994. Cancer incidence in atomic bomb survivors. Part III: Leukemia lymphoma and multiple myeloma, 1950-1987. Radiat Res 137:568-598.

Prosser JS, Edwards AA, Lloyd DC. 1989. A comparison of chromosomal and micronuclear methods for radiation accident dosimetry. In: Goldfinch EP, editor. Radiation protection—theory and practice. Malvern, Pennsylvania: Fourth International Symposium.

Ramalho AT, Nascimento ACH. 1991. The fate of chromosomal aberrations in ^{137}Cs-exposed individuals in the Goiania radiation accident. Health Phys 60:67-70.

Ramalho AT, Nascimento ACH, Natarajan AT. 1988. Dose assessments by cytogenetic analysis in the Goiania (Brazil) radiation accident. Radiation Protection Dosimetry 23:97-100.

Robkin MA. 1992. Experimental release of ^{131}I: the Green Run. Health Phys 62:487-495.

Rogers DR. 1975. Mound Laboratory environmental plutonium study, 1974. Springfield, Virginia: Technical Information Service, U. S. Department of Commerce.

Ron E, Preston DL, Mabuchi I, Thompson DE, Soda M. 1994. Cancer incidence in atomic bomb survivors. Part IV: Comparison of cancer incidence and mortality. Radiat Res 137:598-5112.

Rothman KJ. 1990. A sobering start for the cluster busters' conference. Am J Epidemiol 132:S-6-S-13.

Sehmel G. 1980. Particle and gas dry deposition review: A review. Atmos Environ 14:983-1011.

Shimizu Y, Schull WJ, Kato H. 1990. Cancer risk among atomic bomb survivors. The RERF life span study. JAMA 264:622-623.

Shipler DB, Napier BA. 1994. HEDR modeling approach. PNWD-1983 HEDR Rev. 1. Hanford, Washington: Battele, Pacific Northwest Laboratories.

Shore RE, Iyer V, Altshuler B, Pasternack B. 1992. Use of human data in quantitative risk assessment of carcinogens: Impact on epidemiologic practice and the regulatory process. Regul Toxicol Pharmacol 15:180-221.

Slade DH. 1968. Meteorology and atomic energy. AEC Technical Information Center. TID-24190.

Slovic P. 1987. Perception of risk. Science 236:280-285.

Stevens W, Thomas DC, Lyon JL, Till JE, Kerber RA, Simon SL, Lloyd RD, Elghany NA, Preston-Martin S. 1990. Leukemia in Utah and radioactive fallout from the Nevada Test Site: A case-control study. JAMA 264:585-591.

Stevenson KA, Hardy EP. 1993. Estimate of excess uranium in surface soil surrounding the Feed Materials Production Center using a requalified data base. Health Phys 65:283-287.

Stidley CA, Samet J. 1994. Assessment of ecologic regression in cancer and indoor radon. Am J Epidemiol 139:312-322.

Straume T, Lucas JN. 1993. A comparison of the yields of translocations and dicentrics measured using fluorescence *in situ* hybridization. Int J Radiat Biol 64:185-187.

Straume T, Lucas JN. 1995. Validation studies for monitoring of workers using molecular cytogenetics. In: Biomarkers in occupational health: Progress and perspectives. Mendelsohn ML, Peeters JP, Normandy MJ, editors, Washington, DC: Joseph Henry Press. p. 174-193.

Straume T, Lucas JN, Tucker JD, Bigbee WL, Langlois RG. 1992. Biodosimetry for a radiation worker using multiple assays. Health Phys 62:122-130.

Thompson DE, Mabuchi K, Soda M, Tokunaga M, Ochikubo S, Sugimoto S, Ikeda T, Terasaki M, Izumi S, Preston DL. 1994. Cancer incidence in atomic bomb survivors. Part II: Solid tumors, 1958-1987. Radiat Res 137:S-17–S-67.

Till JE. 1993. Dose reconstruction: Establishing the framework for a public study. Presented at the National Academy of Sciences Workshop on Dose Reconstruction for Epidemiological Uses. October 25-27, 1993. Washington, DC.

Trabalka JR, Auerbach SI. 1990. One western perspective of the 1957 Soviet nuclear accident. In: Proceedings on seminar on comparative assessment of the environmental impact of radionuclides released during three major nuclear accidents: Kyshtym, Windscale, Chernobyl. EUR-13574. p 41-69.

[TSP] Technical Steering Panel. 1990. Technical Steering Panel of the Hanford Environmental Dose Reconstruction Project. Initial Hanford radiation dose estimates. Washington State Department of Ecology.

[TSP] Technical Steering Panel. 1992. Technical Steering Panel of the Hanford Environmental Dose Reconstruction Project. Summary of final sources. Initial Hanford radiation dose estimates. Washington State Department of Ecology.

[UNSCEAR] United Nations Scientific Committee on the Effects of Atomic Radiation. 1988. Sources, effects, and risks of ionizing radiation. New York, NY: United Nations.

[UNSCEAR] United Nations Scientific Committee on the Effects of Atomic Radiation. 1993. Sources and effects of ionizing radiation. UNSCEAR 1993 report to the General Assembly, with scientific annexes. New York, NY: United Nations.

[UNSCEAR] United Nations Scientific Committee on the Effects of Atomic Radiation. 1994. Sources and effects of ionizing radiation. UNSEAR 1994 report to the General Assembly with scientific annexes. New York, NY: United Nations.

[US EPA] United States Environmental Protection Agency. 1994. Estimating radiogenic cancer risks. Document Number EPA 402-R-93-076.

Vital statistics of the United States. 1980. U.S. Department of Health and Human Services. Hyattsville, Maryland: National Center for Health Statistics.

Voillequé PG, Gesell TF. 1990. Evaluation of environmental radiation exposures from nuclear testing in Nevada: A symposium. Health Phys 59:501-502.

Voillequé PG, Meyer KR, Schmidt DW, Killough GG, Moore RE, Ichimura VI, Rope SK, Shleien B, Till JE. 1991. The Fernald Dosimetry Reconstruction Project. Tasks 2 and 3: radionuclide source terms and uncertainties—1960-1962. Neeses, South Carolina Radiological Assessments Corporation.

Wachholz BW. 1990. Overview of the National Cancer Institute's activities related to exposure of the public to fallout from the Nevada Test Site. Health Phys 59:511-514.

Walker A, Blettner M. 1985. Comparing imperfect measures of exposure. Am J Epidemiol 121:783-790.

Whicker FW, Kirchner TB. 1987. PATHWAY: A dynamic food-chain model to predict radionuclide ingestion after fallout deposition. Health Phys 52:717-737.

Whicker FW, Kirchner TB, Breshears DD, Otis MD. 1990. Estimation of radionuclide ingestion: The "PATHWAY" food-chain model. Health Phys 59:645-657.

Wolff S. 1991. Biological dosimetry with cytogenetic end points. In: Gledhill BL, Mauro F, editors. Biological dosimetry and detection of disease-specific chromosome aberrations. New York: Wiley-Liss. p 351-362.

Zvonov IA, Balonov MI. 1993. Radioiodine dosimetry and prediction of consequences of thyroid exposure of the Russian population following the Chernobyl accident. In: Morwin S, editor. The Chernobyl papers. Vol. I: Doses to the Soviet population and early health effects studies. Richland, Washington: Research Enterprises Publishing Segment. p. 71-125.

APPENDIX

A
Representative Dose
Reconstruction Studies

$\mathbb{A}$ NUMBER OF DOSE reconstruction
studies of environmental releases of radioactive materials have been either completed or undertaken, and there are lessons to be learned from these efforts. The most important dose reconstructions have been associated with nuclear weapons testing (Maralinga, Pacific Test Site, Nevada Test Site, etc.); reactor accidents (Chernobyl, Three Mile Island, Windscale); routine releases from installations of the nuclear fuel cycle, especially during the early years of operation (Fernald, Hanford, Techa River); and careless disposal of industrial or medical radioactive sources (Goiania). The manner in which doses were reconstructed in some of these studies is discussed below.

NEVADA TEST SITE

The first modern dose reconstruction study done in the United States involved the Nevada Test Site (Voillequé and Gesell 1990). At this location, roughly 100 above-ground tests of nuclear weapons were conducted in the 1950s; tests also occurred in the early 1960s. The monitoring of the fallout (as determined by measurements of external gamma-exposure rate) from these tests was extensive within the close-in area, and calculations of the external gamma dose were made and tabulated for many communities (Anspaugh and Church 1986, Anspaugh and others 1990). Late in the 1970s, considerable controversy developed surrounding allegations that leukemias, and subsequently other cancers, had been caused by the exposure, and a decision was made to carry out a dose reconstruction study to reevaluate the estimates of external exposure and dose and to attempt for the first time a complete assessment of exposure and dose from the ingestion and inhalation of radionuclides (Church and others 1990).

The decision to perform this study was made in advance of any convincing demonstration that a dose reconstruction could actually be performed. An early debate ensued on the possible options of performing such a study with one option being a "model only" study using existing fallout and atmospheric transport models. The option eventually selected was to make maximum possible use of the existing historic data, which consisted of more than 100,000 measurements of the external gamma-exposure rate (Grossman and Thompson 1990) and more than 10,000 measurements of ground-level concentrations of radionuclides in air (Cederwall and others 1990). Beck and Krey (1983) had demonstrated that doses could be reasonably well reconstructed on the basis of contemporary measurements of ^{137}Cs and $^{239+240}Pu$ deposition densities in undisturbed soil, together with the determination of the ^{240}Pu-to-^{239}Pu ratio. The decision was made to enlarge the geographic coverage of the study to include the entire states of Arizona, Nevada, New Mexico, and Utah, as well as several counties of California, Colorado, Idaho, Oregon, and Wyoming (Church and others 1990).

Two of the more difficult aspects of the study involved determining the relationship between the measured values of the external gamma-exposure rate and the deposition densities of the more than 100 radionuclides released by the tests and to then calculate the intake of different radionuclides by the exposed individuals and population. The first problem was solved with Beck's (1980) calculations of the external gamma-exposure rate per unit deposition density of a given radionuclide and by the development of normalized source terms (per mR hr^{-1} of exposure rate 12 hr after detonation) (Hicks 1982, 1990). (The mR represents a unit of exposure called milliroentgens.) The second problem was solved by the development of a dynamic, seasonally dependent food chain model with specific consideration of uncertainty (Whicker and Kirchner 1987, Whicker and others 1990).

Example results (Henderson and Smale 1990, Ng and others 1990) of this study have been available for some time and intermediate results (Beck and Anspaugh 1990) have been used in epidemiologic studies (Kerber and others 1993, Stevens and others 1990) and to assess the thyroid doses of ^{131}I received by the populations of the contiguous United States (Bouville and others 1990, Wachholz 1990). One important aspect of the Nevada Test Site study was the extensive use of contemporary measurements of long-lived materials in soil to confirm and extend the original measurements. Furthermore, a convincing confirmation of the essential correctness of the external dose calculations was provided by contemporary solid-state dosimetry measurements of the thermoluminescence signals obtained from quartz contained in bricks in homes in the downwind areas (Haskell and others 1994).

CHERNOBYL

The April 1986 accident at Unit 4 of the Chernobyl nuclear power plant, in the Ukraine, about 30 km south of the border with Belarus, was the most severe in the history of the nuclear industry. The accident caused the death of 31 power plant employees and firemen from acute radiation exposures and burns. It brought about the early evacuation of 135,000 people and resulted in the contamination of vast regions of Belarus, Russia, and the Ukraine, as well as, to a lesser extent, many countries of northern Europe. About 330 petabecquerels[1] (PBq) (8.91×10^6 Ci) of ^{131}I, 35 PBq (9.45×10^5 Ci) of ^{134}Cs, and 70 PBq (1.89×10^6 Ci) of ^{137}Cs were released into the atmosphere over a period of 10 d. During that time, the winds blew in many directions, so that the radioactive materials released were transported over different regions. Most of the materials in the radioactive cloud were deposited on the ground, largely through precipitation, and this resulted in the contamination of milk and other foodstuffs.

The short-lived ^{131}I caused high thyroid exposures, especially among children, in the first few weeks after the accident. Radioiodine concentrations were measured in about 250,000 persons in Belarus, 150,000 persons in Ukraine, and 30,000 persons in Russia. Preliminary estimates indicate that the thyroid doses received by children range up to 30 Gy (3,000 rad) or more. A preliminary thyroid dose distribution among children in the most heavily contaminated districts of Belarus who were under the age of 7 years at the time of the accident is provided in Table A-1. The estimated arithmetic mean thyroid doses in subgroups of the same populations vary from 0.21 to 1.06 Gy or 21–106 rad (Table A-2). Efforts are being made to reconstruct the individual thyroid doses received by the most exposed populations of Belarus (Gavrilin and others 1992), Russia (Zvonov and Balonov 1993), and Ukraine (Likhtarev and others 1993), on the basis of the thyroid measurements and of personal interviews on dietary and lifestyle habits. Serious technical difficulties have been encountered when the thyroid measurements were carried out by inexperienced individuals using instrumentation that was not specifically designed for this type of measurement.

In addition, the longer lived ^{134}Cs, and, more important, ^{137}Cs, deliver doses to the entire body and will be present in the environment for de-

TABLE A-1 Preliminary Thyroid Dose Distribution Among Children in the Most Heavily Contaminated Districts of Belarus Who Were Less Than 7 Years Old at the Time of the Accident (Gavrilin and others 1992)

Thyroid Dose Range (Gy)	Nine Districts of Gomel Region, 32,420 Children		Five Districts of Mogilev Region, 14,240 Children		Fourteen Districts of Gomel and Mogilev, 46,660 Children	
	Number	%	Number	%	Number	%
0.00-0.03	15,128	46.66	9,637	67.68	24,675	53.08
0.03-0.75	8,951	27.61	2,975	20.88	11,926	23.55
0.75-2.00	4,924	15.18	1,345	9.45	6,269	13.44
2.00-5.00	2,428	7.49	251	1.76	2,679	5.74
5.00-10.0	693	2.14	28	0.20	721	1.55
10.0-20.0	274	0.85	4	0.03	278	0.60
20.0-30.0	20	0.06			20	0.04
30.0-40.0	2	0.01			2	<0.01

TABLE A-2 Estimated Arithmetic Mean Thyroid Doses to Children Under Age 7 in the Most Contaminated Districts of Belarus (Gavrilin and others 1992)

Oblast (Province)	Number of Districts	Population Type	Population Size	Mean Thyroid Dose (Gy)
Gomel	9	Rural	23,900	1.06
		Urban	8,600	0.44
Mogilev	5	Rural	9,300	0.44
		Urban	4,900	0.21
Gomel and Mogilev	14	Rural	33,200	0.88
		Urban	13,500	0.36

cades to come. The resulting doses, which are less important than those delivered by ^{131}I, also are being reconstructed for the populations living in contaminated areas. Doses from external irradiation are best determined from transport calculations based on measured cesium concentrations in soil and on thermoluminescent dosimeter (TLD) measurements, whereas the doses from internal irradiation are estimated from whole-body counting data or from ^{137}Cs concentrations in milk associated with information on milk consumption rates.

Attention is also being paid to [90]Sr. The migration of this radionuclide from the soil through the terrestrial food chain could result in later years in doses from internal irradiation similar to those from [137]Cs.

Finally, the resuspension into ground-level air of ^{241}Am, ^{239}Pu, and other long-lived transuranics is likely to lead to exposure to future generations, when ^{137}Cs and ^{90}Sr have decayed to negligible values.

THREE MILE ISLAND

On March 28, 1979, the No. 2 Unit of the Three Mile Island reactor in Middletown, Pennsylvania, had a loss-of-coolant accident that led to a partial core melt-down and the subsequent release of approximately 400 PBq (10 megacuries[2] [MCi]) of noble gases. The exact magnitude of this release is not known because the effluent monitors were insufficient to monitor the amount of the release. Despite the large release of noble gases, the amount of radioiodine released was estimated to be only 0.4-1 terabecquerel[3] (TBq) (10-30 Ci). The iodine release was estimated from measurements performed on charcoal absorbers in the effluent line. Because the accident occurred in the early spring, cows were not yet in pastures and the amount of radioiodine transferred to milk was low, leading to maximum concentrations in milk of about 1.5 Bq L^{-1} (40 picocuries[4] L^{-1}).

Several approaches were considered for estimating individual and population doses from the noble gases. The most useful was to rely on actual measurements of doses from external irradiation made by TLDs surrounding the site.

The maximum individual dose was estimated to be less than 1 millisievert (mSv; 100 mrem) and the collective dose was 20-35 person-sievert (2,000-3,500 person-rem). The average dose to the 2 million people residing within 50 miles was 0.015 mSv (1.5 mrem).

The Three Mile Island case illustrates the importance of using environmental measurement data for dose reconstruction when an accurate source term cannot be determined because of instrumental deficiencies or other causes.

FERNALD

The Feed Materials Production Center (FMPC) located at Fernald, Ohio, was, between 1951 and 1989, a government-owned, contractor-operated facility for producing uranium metal products used as feed materials in the production of nuclear weapons. Environmental releases of radioactive materials consisted mainly of uranium, thorium, and radon into the atmosphere.

An environmental dose reconstruction is being conducted at Fernald. From preliminary information, it seems that the principal pathway is inhalation of uranium. The main difficulty so far encountered in the environmental pathway analysis appears to be that there are not enough local meteorologic data for 1951 to 1986. A recording meteorologic tower has been in regular operation at the site only since 1986. Data for earlier times (from 1951 to 1986) are available for regional airports, in Cincinnati and Dayton, but those sites are more than 50 km from Fernald. The options available to the dose reconstruction team were to use recent Fernald data for the earlier period, with a large degree of uncertainty applied to air concentration estimates, or to use a surrogate historic data set based on regional airport data. After examination of the meteorologic data available, it was concluded that the air dispersion predictions for the dose reconstruction study should be based on hourly wind and stability data from the Fernald meteorologic tower, with uncertainties based partly on the relationships between past and recent Cincinnati airport data.

HANFORD

The nuclear weapons facility at Hanford, Washington, released substantial quantities of radioactive materials into the atmosphere and the Columbia River from its plutonium production reactors and fuel reprocessing facilities. Two plutonium production reactors started operating at Hanford in December 1944. Two fuel reprocessing plants began extracting plutonium in the same month, and a third production reactor was added in 1945 (Cate and others 1990). The bulk of the releases of radioactive materials—^{131}I discharged from the fuel-reprocessing plants—into the atmosphere occurred between 1944 and 1947. The amount of ^{131}I released during that period is estimated at 685 kilocuries (kCi) (25 PBq) on the basis of the quantity and origin of reactor fuel reprocessed and on the time interval between removal from the reactors and reprocessing (Heeb 1992, TSP 1992, Robkin 1992). The production processes also resulted in the release of other radioactive materials to the atmosphere, to the Columbia River, and to groundwater.

A dose reconstruction study began in 1988. In the first phase, scientists developed and tested methods for reconstructing the radiation doses to people who lived in the 10 Washington and Oregon counties closest to Hanford. To do this, they focused on the atmospheric releases of ^{131}I from 1944 to 1947 and on the releases to the Columbia River from 1964 to 1966. Phase I of the study was completed in 1990 (TSP 1990). According to the preliminary results, the highest doses from ^{131}I released to the atmosphere were to infants and young children drinking milk from cows pastured in north Franklin County. Thyroid doses for most individuals in this group of about 1,400 range from 0.15 to 6.5 Gy (15 + 650 rad). Individual doses from radioactive materials released to the Columbia River during 1964 through 1966 are estimated to have been much lower than doses from contaminated milk during the 1940s (TSP 1990). The final report on the study was published in 1994 (Shipler and Napier 1994).

The main difficulty encountered in the part of the dose reconstruction study related to the atmospheric releases of ^{131}I is that very few measurements of environmental radioactivity were made in the mid-1940s, so that the study is based largely on the use of models. In addition, important data about the commercial distribution of milk were not available and had to be obtained from interviews with people who remembered the activities of local dairies in the 1940s.

An important characteristic of this dose reconstruction study is the extent to which the public is being kept informed on the progress of the work. The Technical Steering Panel (TSP) of the Hanford Environmental Dose Reconstruction Project, the group responsible for the study, consists of experts in various fields, as well as representatives of the Oregon and Washington state governments, of regional groups of Native Americans, and of members of the public. The TSP made all meetings open to the public and declared that any documents it received were available to the public and media. In addition, the TSP communicates to the public in quarterly newsletters, fact sheets, a video, and a toll-free telephone line.

TECHA RIVER

The Chelyabinsk-40 center, located near the town of Kyshtym in Russia, was the first Soviet nuclear installation dedicated to the production of plutonium for military purposes (UNSCEAR 1993). A uranium-graphite reactor with an open cooling-water system was commissioned in June 1948, and a fuel-reprocessing plant began operating in December 1948 (Nikipelov and others 1990). Liquid releases to the Techa River from 1949 to 1956 amounted to 2.7 MCi (100 PBq); 95% of this release was discharged between March 1950 and November 1951 (Kossenko 1991). The main constituents released were ^{89}Sr (8.8%), ^{90}Sr (11.6%), ^{137}Cs (12.2%), rare-earth isotopes (26.8%), ^{95}Zr-^{95}Nb (13.6%), and ruthenium isotopes (25.9%). These large releases appear to have resulted primarily from a lack of waste treatment capability and from the storage of radioactive wastes in open, unlined earthen reservoirs (Trabalka and Auerbach 1990). A hydrologic isolation system, including a small reservoir called Lake Karachay, was built after 1952 to contain the low- and intermediate-level wastes (UNSCEAR 1993).

The population along the Techa River was exposed to both external and internal irradiation. External irradiation was caused by gamma radiation from ^{137}Cs, ^{106}Ru, and ^{95}Zr-^{95}Nb in the flood plains, in vegetable gardens near houses, and inside houses. Internal irradiation mainly resulted from consumption of water and of local foodstuffs contaminated with ^{89}Sr and [90]Sr.

Average cumulative effective doses are estimated to have been as high as 1.4 Sv in the village of Metlino, 7 km downstream from the point of discharge. The evacuation of the village started in 1953; from 1955 to 1960, inhabitants of another 19 settlements were moved away from the river. Altogether, 7,500 persons were relocated (Kossenko 1991). More recently, Kossenko and Degteva (1994) have estimated that the average effective dose to their Tartar-Bashkir population of 6,123 persons is 0.37 Sv (37 rem); that to the Russian population of 20,563 is 0.13 Sv (13 rem). The person-years weighted mean doses to the bone marrow and soft tissue have been estimated to be 0.37 and 0.14 Gy, respectively (37 and 14 rad).

Scientists from the Ural Research Center for Radiation Medicine are following up 28,000 persons who lived in 38 villages along the banks of the Techa River in 1949. External doses have been reconstructed on the basis of gamma-dose rate measurements in the early 1950s along the river bank, on the shore within a few hundred meters of the river, in specified areas of villages, and inside houses. A survey was also made of the lifestyle habits of the populations from the riverside villages (Kossenko and others 1992). Internal irradiation doses have been reconstructed from measurements of the surface-beta activity of teeth, done from 1960 to 1976, and from whole-body counter measurements begun in 1974 (Kossenko and others 1992).

GOIANIA

In 1987, a ^{137}Cs teletherapy (radiation therapy) source in Goiana, Brazil, was broken by a scrap metal collector who dispersed parts of the 50.9 TBq (1,375 Ci) source (cesium chloride powder) in his house and garden and to other properties in this city of 1.3 million inhabitants. The accident has been described by several papers in a special issue of *Health Physics* (1991). External gamma irradiation was the main cause of radiation exposure. Acute radiation sickness led to the death of 4 persons whose doses were reconstructed largely on the basis of hematologic observations (Brandão-Mello and others 1991). Dose reconstruction for other persons was begun in 1987 (more than 100,000 were considered to be potentially affected) and used whole-body counting where possible, dose rate measurements in front of highly contaminated persons for which whole-body counter measurements were not possible because of too high a count rate, and dicentric chromosome biologic dosimetry (Ramalho and others 1988). External exposure fields were reconstructed with hand-held dose rate meters (several weeks later and partially after the decontamination most urgently needed) and TLD measurements were made of wall materials. Time-and-space information was assembled on the whereabouts of individuals during the time of the accident. Internal doses of ^{137}Cs were reconstructed by assessment of ^{137}Cs concentrations in air and in food.

The following pitfalls were encountered in the dose reconstruction: The measured dose rates were influenced by ^{137}Cs in the environment as well as on the skin and clothes of the persons studied. Estimating organ doses from the measured values is difficult. The external skin contamination of the persons also influenced the interpretation of whole-body counting measurements. Often, the count rate was too high for these sensitive instruments. Whole-body counting cannot indicate the external exposures that contributed most of the doses. There is no calibration curve for the induction of dicentric chromosomes in human lymphocytes by combined internal and external ^{137}Cs irradiation at low doses, and sometimes for partial-body irradiation (a clear research need); in addition, the number of people potentially affected was prohibitively high (more than 100,000 individuals were at risk). The shielding by structures in the urban environment made retrospective determination of external gamma fields difficult and uncertain. The highly heterogeneous nature of the contamination and of its resulting radiation field required an accuracy in the retrospective time-and-space records that could not be met in most cases. The early environmental dispersion of ^{137}Cs resulted from human actions and from resuspension that could not be reconstructed with sufficient detail. Detailed measurements could not be carried out until several weeks after the accident, when inhalation and ingestion played a negligible part in the total exposure and uptake.

In summary, the reconstruction of the doses of highly affected people was carried out by clinical evaluation, dose reconstruction by biologic dosimetry was severely hampered by the lack of calibration curves for combined external and internal ^{137}Cs irradiation at a low dose rate, and case-specific exposure pathway analysis models (necessarily unvalidated) based on measured dose rates and ^{137}Cs concentrations weeks later were only useful for approximative retrospective dose estimates for reference persons.

NOTES

1. A petabecquerel (PBq) is 10^{15} Bq.
2. A megacurie (MCi) is 10^6 Ci.
3. A terabecquerel (TBq) is 10^{12} Bq.
4. A picocurie (pCi) is 10^{-12} Ci.

APPENDIX
B
Workshop Agenda

DOSE RECONSTRUCTION FOR EPIDEMIOLOGIC USES
AGENDA
Lecture Room, National Academy of Sciences
2101 Constitution Avenue, NW
Washington, DC

Sunday, October 24, 1993
5:30/6:30   RECEPTION (Rotunda)/DINNER (Members Room)

Monday, October 25, 1993
8:00 - 8:30   Continental Breakfast
8:30 - 8:40   Welcome

John Zimbrick (Director, Board on Radiation Effects Research, National Research Council)

James Smith (Chief, Radiation Studies Branch, Centers for Disease Control)

8:40 - 9:00   **Introduction**—William Schull (Chair, Committee on an Assessment of CDC Radiation Studies)

9:00 - 10:00   **Dose Reconstruction**—John Till
(Radiological Assessments Corporation, Neeses, SC)

10:00 - 10:45   **Source Term Estimation**—Paul Voillequé
(MJP Risk Assessment, Inc., Idaho Falls, ID)

10:45 - 11:00   **BREAK**

11:00 - 11:45   **Environmental Pathways: Models and Approaches**   Owen Hoffman (Center for Risk Analysis, Oak Ridge, TN)

11:45 - 12:30   **Biological Dosimetry**—Sheldon Wolff
(University of California, San Francisco, CA)

12:30 - 1:30	**LUNCH**

1:30 - 2:15	Radiation Dose Assessment - Bruce Napier (University of Washington, Seattle, WA)

2:15 - 3:00	When Is an Epidemiological Study Appropriate? Genevieve Matanoski (The Johns Hopkins University, Baltimore, MD)

3:00 - 3:15	Charge to Working Groups—William Schull

3:15 - 5:30	Break-out Sessions

Environmental Pathways	Herwig Paretzke (Institut fur Strahlenschutz, Neuherberg, Germany)
	André Bouville (National Cancer Institute, Bethesda, MD)
Source Term Estimation	James Martin (University of Michigan, Ann Arbor, MI)
Epidemiology	Roy Shore (New York University Medical Center, New York, NY)
	Leeka Kheifets (Electric Power Research Institute, Palo Alto, CA)
Biomarkers	Richard Albertini (University of Vermont, Burlington, VT)
	Stephen A. Benjamin (Colorado State University, Ft. Collins, CO)
Radiation Dose Assessment	Chris Nelson (Environmental Protection Agency, Washington, DC)
	Robert Thomas (Argonne National Laboratory, Argonne, IL)

Tuesday, October 26, 1993

8:00 - 8:30	Continental Breakfast
8:30 - 12:30	Break-out Sessions
12:30 - 1:30	LUNCH
1:30 - 2:30	Assembly of Participants
2:30 - 4:30	Break-out Sessions
4:30 - 5:30	Assembly of Participants

Wednesday, October 27, 1993

8:00 - 8:30	Continental Breakfast
8:30 - 10:30	Status of Draft Document (Report by Break-out Session Leaders)
10:30 - 12:30	Break-out Sessions
12:30 - 1:30	LUNCH
1:30 - 3:30	Break-out Sessions
3:30 - 5:00	Workshop Summary (Distribution of Draft Document)

APPENDIX
C
Workshop Participants

KEYNOTE SPEAKERS

Owen Hoffman
Senes Oak Ridge, Inc.
Center for Risk Analysis
677 Emory Valley Road, Suite C
Oak Ridge, TN 37830
(615) 483-6111
(615) 481-0060 (fax)

ENVIRONMENTAL PATHWAYS:
MODELS AND APPROACHES

Genevieve Matanoski
School of Hygiene and Public Health
The Johns Hopkins University
615 North Wolfe Street
Baltimore, MD 21205
(410) 955-8183
(410) 276-0290 (fax)

WHEN IS AN EPIDEMIOLOGICAL
STUDY APPROPRIATE?

Bruce Napier
Battelle, Pacific Northwest Laboratories
P.O. Box 999
Richland, WA 99352
(509) 375-3896
(509) 375-2019 (fax)

RADIATION DOSE ASSESSMENT

John Till   DOSE RECONSTRUCTION
Radiological Assessment Corporation
Route 2, Box 122
Neeses, SC 29107
(803) 536-4883
(803) 534-1995 (fax)

Paul Voillequé   SOURCE TERM ESTIMATION
MJP Risk Assessment, Inc.
P.O. Box 50430
Idaho Falls, ID 83405-0430
(208) 529-9171
(208) 529-3795 (fax)

Sheldon Wolff   BIOLOGICAL DOSIMETRY
Laboratory of Radiobiology and
 Environmental Health
P.O. Box 0750
University of California, San Francisco
San Fracisco, CA 94143
(415) 476-1636
(415) 476-0721 (fax)

SOURCE TERM ESTIMATION

Sanford C. Cohen
1355 Beverly Road
Suite 250
McLean, VA 22101
(703) 893-6600
(703) 821-8236 (fax)

Geoffrey G. Eichholz
Department of Nuclear Engineering
Georgia Institute of Technology
Atlanta, GA 30332-0225
(404) 894-3207
(404) 894-3733 (fax)

J. Charles Jennett
Clemson University
207 Sikes Hall
Clemson, SC 29634
(803) 656-3243
(803) 656-0851 (fax)

George Kerr
Oak Ridge National Laboratory
P.O. Box 2008
Oak Ridge, TN 37831-6383
(615) 574-6258
(615) 574-1778 (fax)

James E. Martin
School of Public Health
Department of Environmental and Industrial Health
University of Michigan
109 Observatory Street
Ann Arbor, MI 48109-2029
(313) 936-0763
(313) 764-9424 (fax)

Allan Richardson
Office of Radiation and Indoor Air (6602J)
U.S. Environmental Protection Agency
Washington, DC 20460
(202) 233-9290
(202) 233-9629 (fax)

George Sherwood
Department of Energy
NE-80
Washington, DC 20585
(301) 903-4162
(301) 903-7738 (fax)

Paul Voille99é
MJP Risk Assessment, Inc.
P.O. Box 50430
Idaho Falls, ID 83405-0430
(208) 529-9171
(208) 529-3795 (fax)

ENVIRONMENTAL PATHWAYS

Lynn R. Anspaugh
Lawrence Livermore National Laboratory
7000 East Avenue
Livermore, CA 94550
(510) 422-3820
(510) 423-6785 (fax)

André Bouville
Radiation Effects Branch
National Cancer Institute (EPN 530)
Bethesda, MD 20892
(301) 496-9326
(301) 496-1224 (fax)

Owen Hoffman
Center for Risk Analysis
Senes Oak Ridge Inc.
677 Emory Valley Road, Suite C
Oak Ridge, TN 37830
(615) 483-6111
(615) 481-0060 (fax)

Il'ya A. Likhtarev
Ukrainian Scientific Center of Radiation
  Medicine
Melnicova Street 53
Kiev 252050
Ukraine
011-7-044-213-7192

Herwig G. Paretzke
GSF—Forschungszentrum fur Umwelt und
  Gesundheit
Institut fur Strahlenschutz
Ingolstadter Landstrasse
D-8042 Neuherberg, Germany
011-49-89-3187-4006
011-49-89-3187-3323 (fax)

Harold T. Peterson
Department of Energy
Office of Environmental Guidance
Air, Water, and Radiation Division
EH-232, Room GA-098
1000 Independence Avenue, SW
Washington, DC 20585
(202) 586-9640
(202) 586-3915 (fax)
(202) 586-8134 (fax)

Bruce Wachholz
Radiation Effects Branch
National Cancer Institute
EPN 530
Bethesda, MD 20892
(301) 496-9326
(301) 496-1224 (fax)

BIOMARKERS

Mitoshi Akiyama
Department of Radiobiology
Radiation Effects Research Foundation
5-2 Hijiyama Park, Minami-ku
Hiroshima City 732
Japan
011-81-82-261-3131
011-81-82-263-7279 (fax)

Richard J. Albertini
VRCC Genetics Laboratory
University of Vermont College of Medicine
32 N. Prospect Street
Burlington, VT 05401
(802) 863-5716
(802) 656-8788 (fax)

Stephen A. Benjamin
Center of Environmental Toxicology
College of Veterinary Medicine and Biomedical Sciences
Colorado State University
Ft. Collins, CO 80523
(303) 491-8522
(303) 491-8304 (fax)

Antone L. Brooks
Battelle, Pacific Northwest Laboratories
P.O. Box 999, P7-50
Richland, WA 993352
(509) 376-4487
(509) 376-0302 (fax)

Ronald Jensen
Department of Laboratory Medicine
University of California
Division of Molecular Cytometry
MCB 230
San Francisco, CA 94143-0808
(415) 476-3383
(415) 476-8218 (fax)

Albrecht Kellerer
Institut fur Strahlen biologie
Ludwig - Maximilians - U. Munchen
Schillerstr. 42/Munich D-8000
Germany
011-004-9-89-5996-818
011-004-9-89-5996-840 (fax)

Tore Straume
Dosimetry and Dose Response Group
Lawrence Livermore National Laboratory
P.O. Box 808
Livermore, CA 94551
(510) 422-5138
(510) 422-5748 (fax)

Richard Wilson
Department of Physics
Lyman Building, Room 231
Harvard University
Cambridge, MA 02138
(617) 495-3387
(617) 495-0416 (fax)

Sheldon Wolff
Laboratory of Radiobiology and
  Environmental Health
P.O. Box 0750
University of California, San Francisco
San Francisco, CA 94143
(415) 476-1636
(415) 476-0721 (fax)

RADIATION DOSE ASSESSMENT

Mikhail Balanov
Institute of Radiation Hygiene
19101 St. Petersburg
Russia
011-7-812-233-4843
011-7-812-315-1701 (fax)

Robert Catlin
University of Texas Health Science Center
SHI R-108
1343 Moursund Street
Houston, TX 77030
(713) 794-1764
(713) 792-2315 (fax)

Stephanie Haywood
National Radiological Protection Board
Chilton, Didcot
Oxfordshire OX11 ORQ
England
011-44-235-831-600
011-44-235-833-891 (fax)

Hans Menzel
Radiation Protection Research Unit
Commission of the European Communities
DG XII-D-3 ARTS 3/1
200 Rue de la Loi 1049
Brussels, Belgium
011-32-2-295-4045
011-32-2-296-6256 (fax)

Bruce Napier
Battelle, Pacific Northwest Laboratories
P.O. Box 999
Richland, WA 99352
(509) 375-3896
(509) 375-2019 (fax)

Christopher B. Nelson
Office of Radiation & Indoor Air (6602J)
U.S. Environmental Protection Agency
401 M Street, SW
Washington, DC 20460
(202) 233-9209
(202) 233-9629 (fax)

Henry D. Royal
Division of Nuclear Medicine
Mallinckrodt Institute of Radiology
Washington University Medical
  Center
510 South Kingshighway Boulevard
St. Louis, MO 63110-1076
(314) 362-2812
(314) 362-2806 (fax)

Robert G. Thomas
495 Windsor Drive
Bigfork, MT 59911
(406) 837-3758
(406) 837-3759

John Till
Radiological Assessments Corporation
Route 2, Box 122
Neeses, SC 29107
(803) 536-4883
(803) 534-1995 (fax)

Henry N. Wagner
Division of Nuclear Medicine & Radiation Health Sciences
The Johns Hopkins Medical Institutes
615 North Wolfe Street, Room 2001
Baltimore, MD 21205-2179
(410) 955-3350
(410) 955-6222 (fax)

EPIDEMIOLOGY

Ethel S. Gilbert
Battelle, Pacific Northwest Laboratories
P.O. Box 999
Battelle Boulevard
Richland, WA 99352
(509) 376-7347
(509) 376-4533 (fax)

Seymour Jablon
National Cancer Institute
Radiation Epidemiology Branch
6130 Executive Plaza North
Room 408
Bethesda, MD 20892
(301) 496-6600
(301) 496-0207 (fax)

Leeka I. Kheifets
Electric Power Research Institute
3412 Hillview Avenue
P.O. Box 10412
Palo Alto, CA 94303
(415) 855-8976
(415) 855-1069 (fax)

Genevieve Matanoski
School of Hygiene & Public Health
The Johns Hopkins University
Baltimore, MD 21205
(410) 955-8183
(410) 276-0290 (fax)

Leonard Sagan
EMF Communications Division
Electric Power Research Institute
3412 Hillview Avenue
Palo Alto, CA 94303
(415) 855-2585
(415) 856-6621 (fax)

William J. Schull
Center for Demographic and Population
  Genetics
School of Public Health
University of Texas
6901 Bertner Street
Houston, TX 72225
(713) 792-4685
(713) 792-4615 (fax)

Roy E. Shore
Department of Environmental
  Medicine
New York University Medical Center
Room 204
341 E. 25th Street
New York, NY 10010-2598
(212) 263-6498/6500
(212) 263-8570 (fax)

Glossary

Association: A relationship, generally demonstrated by statistical tests, between an exposure and a health effect. It does not necessarily imply cause and effect.

Ataxia telangiectasia: An inherited disorder associated with an increased risk of cancer, especially lymphoma, and characterized by immunologic, chromosomal, and DNA defects.

Background radiation: The amount of ionizing radiation to which a person is exposed from natural sources, such as terrestrial radiation due to naturally occurring radionuclides in the soil, cosmic radiation originating in outer space, and naturally occurring radionuclides deposited in the body.

Becquerel (Bq): The international unit of activity. One becquerel corresponds to 1 disintegration per second, or 2.7×10^{-11} curies (Ci). Under the international system of nomenclature, becquerels are expressed in multiples of 1 thousand. Thus, 1 thousand becquerels, a kilobecquerel, is abbreviated as KBq, 1 million, a megabecquerel, is abbreviated as MBq, and 1 million billion, a petabecquerel, is abbreviated as PBq.

Bias: Any process in any stage of the collection or analysis of data that tends to produce results that differ systematically from the "true value" of the population variables under study (such as disease rates). In epidemiology the term does not refer to an opinion or point of view.

Biologic marker of effect: A biologic change that is specifically associated with the development of a disease and detectable before the disease is evident.

Biologic marker of exposure or dose: Biologic changes that are specifically induced (in this case) by ionizing radiation that can be measured before any health consequences from exposure are evident, and that can be used to quantify radiation dose.

Biologic marker of susceptibility: A biologic change that demonstrates a differential susceptibility of specific individuals to genotoxicity from ionizing radiation.

Cancer: A general term applied to a variety of diseases characterized by abnormal new growth of tissue and by the spread of that tissue to new sites in the body.

Carcinogen: A substance that causes cancer.

Case: In epidemiology, a person identified as having a particular health end point (such as a specific disease) under investigation.

Case-control study: An epidemiologic investigation that compares exposures in persons who have (cases) or have not (controls) developed the disease under study.

Cluster: A series of cases that occur close together in time, location, or both.   Normally used to describe a grouping of relatively rare diseases, such as leukemia.

Cohort: In epidemiology, a group of persons who are initially free of the disease in question but who have been exposed to the agent under study.   The group is followed up, or traced, after a period of time to quantify the occurrence of the disease in the cohort.

Confidence interval: A range of values bracketing a relative risk or odds ratio estimate calculated in such a way that the range has a specified probability (usually 95%) of including the true, but unknown, value of the risk.   The end points of the confidence interval are called the confidence limits.

Confidence limit: See *confidence interval.*

Confounding: A situation in which an observed association between an exposure and a disease is influenced by other variables associated with the exposure that affect the occurrence of disease.

Confounding variable (confounder): A variable that could explain an observed association (or lack of an association) in an epidemiologic study between an exposure and a disease.   A confounder can create a spurious association between an exposure and a disease or it can mask, weaken, or exaggerate a real association.   Confounding must be ruled out before confidence can be placed in any observed association.

Control: In case-control studies, a person who has not developed the disease of interest and whose exposure is compared with the exposure of those who have. See *case-control study.*

Directed study: Focused selection and analysis of data based on a predetermined method for reconstruction of a source term.

Dose dose-rate effectiveness factor (DDREF): A factor by which the effect caused by a specific dose or dose rate of radiation changes at low as compared to high doses or dose rates.

Dose validation: The use of direct measurement of radionuclide content in the body, dose in environmental samples, individual dose measurements (physical or biodosimetric) to test the results of model evaluations. If the result of direct measurement is different from model prediction, priority must be given to the measurement.

Episodic release: Release of radiation to the environment that is at least 10 times greater than the average amount and for a duration of no more than 10 days.

FISH (fluorescence in situ hybridization): The use of DNA libraries derived specifically from particular chromosomes and conjugated with fluorescent molecules to generate reagents that cause distinctive fluorescence on individual chromosomes. Chromosomal aberrations involving the transfer of DNA from one chromosome to another (such as reciprocal translocations) can be detected using this "chromosome painting."

Follow-up: The process in which epidemiologists track study subjects to observe variables of interest, such as the occurrence of a specific disease, over time.

Genotoxicity: Damage to cellular DNA.

Gray (Gy): The international unit for absorbed dose. One gray is equal to 1 joule per kilogram, or 100 rad; therefore, 10 mGy = 1 rad.

Latency: The period between exposure to a disease-causing agent and the appearance of symptoms. After exposure to ionizing radiation, for example, there is an average latency of 5 years before leukemia develops, and more than 20 years before some other malignant conditions develop.

Leukemia: A disease characterized by rapid and abnormal proliferation of white blood cells in the blood-forming organs (bone marrow, spleen, lymph nodes) and by the presence of immature white blood cells in the peripheral circulation.

Localization of dose: The evaluation of organ doses or anatomic site-specific doses appropriate to the biologic effect for epidemiologic purposes. For example, the dose from iodine radionuclides to the thyroid is necessary for the study of thyroid cancers.

Matched control: In a case-control study, one of a group of persons selected for attributes that are similar to those of persons in the case group. Cases and controls are matched for age, gender, race, or socioeconomic status, for example. See *case-control study*.

Matching variable: A characteristic, such as age or gender, used to select controls. See *matched controls*.

Misclassification error: The erroneous classification of a person into a category. In an epidemiologic study of EMF exposure based on job, for example, including some electricians in the "exposed" group might result in misclassification error if those electricians routinely work on dead circuits.

Mortality: Death; the number of deaths in a given time or place; the death rate.

Mutant: A cell that has been identified as containing DNA with a mutation.

Mutation: A change in DNA sequence.

Mutation spectra: A description of the different kinds of DNA damage (such as deletions, frame shifts, base substitutions, and inversions) that occur when cells are exposed to mutagenic events.

Person-years at risk: A number used as the denominator in incidence and mortality rate calculations; the sum of the years that the persons in the study were observed to see whether they develop the disease or condition of interest. Each person contributes only as many years of observation to the study as he or she is actually observed; if a person leaves, contracts the disease under study, or dies after one year, he or she contributes 1 person-year; if a person leaves after 10 years, he or she contributes 10 person-years.

Population: All inhabitants of a given area.

Power (statistical): In epidemiology, the probability of concluding that an association between an exposure and a disease exists, when the association does not, in fact, exist.

Precision: The closeness of repeated measurements of the same quantity.

Proprietary information: Information protected from public disclosure by ownership rights.

P-value: See *statistical significance*.

Quality of radiation: Epidemiologic analysis must account for the quality of radiation. Dose assessment must be made for each group of radiations with different quality (beta plus gamma radiation, alpha radiation)

rad: The unit of radiation absorbed dose, a traditional derived unit defined as the absorption of 100 ergs/gram.

Rate: In a population, the number of times a specific event occurs during a specific period.

Reference group: A group with which a population under study is compared.

Registry: A file of data on all instances of a particular disease in a population, such as all cancer cases in Iowa. With this information, epidemiologists can calculate incidence rates for other groups.

Relative risk: A measure of risk based on disease rate or death rate that is used frequently in cohort studies. Relative risk indicates the increased or decreased degree of risk among exposed subjects compared with unexposed persons. A relative risk of 1 indicates no association between the exposure and the disease. A relative risk of 2 indicates that the exposed group is twice as likely as the unexposed group to experience the health effect being studied.

Release: A discharge into the environment of radioactive materials either as a result of an accident or in the course of production.

Reliability: The degree to which the results of a study can be replicated. Lack of reliability can arise from divergences between observers or instruments of measurement or from the instability of the attribute being measured.

rem: The traditional derived unit of dose equivalence equal to the dose in rad multiplied by the quality factor (Q) of the radiation. For x-rays and γ rays the Q is usually 1 meaning that an exposure to 1 rad is a rem.

Retinoblastoma: A malignant embryonic neoplasm of the retina of the eye.

Retrospective cohort study: An epidemiologic study that follows a cohort from some time in the past to a more recent time in the past. Existing records, such as occupational records or community residence records, are generally used to identify groups for study.

Risk: The probability that an event will occur, such as the probability that an individual will become ill or die within a stated period.

Risk factor: An aspect of personal behavior or lifestyle, an environmental exposure, or an inborn or inherited characteristic that is known from epidemiologic evidence to be associated with adverse health effects.

Sample size: The number of people selected (sampled) from a population to be the subjects of an epidemiologic study.

Scoping study: Use of basic information about a site to provide bounding estimate for initial decisions on conducting an epidemiologic investigation.

Security information: Information protected from public disclosure for reasons of national security, such as information about the design of nuclear weapons.

Selection bias: Error that arises from systematic differences in characteristics between those who have been exposed to different amounts of a substance (in a cohort study), or between those who have and have not developed the disease of interest (in a case-control study). For example, selection bias would exist in a case-control study of radiation and lung cancer that did not account for persons in the case group who tended to smoke more cigarettes than did the controls.

Sensitivity analysis: IEevaluation of the extent to which changes in the values of independent variables (or model parameters) of an equation (or mathematical model) bring about changes in the model result. Within the context of an uncertainty analysis, it is the evaluation of the extent to which uncertainty in the parameters (and in the functional relationships) in a mathematical model contributes to the overall uncertainty in the model result. By identifying the terms that dominate the overall uncertainty in the model result, a sensitivity analysis is an important tool for guiding research efforts.

Sievert (Sv):   The SI unit of dose equivalence equal to the dose in grays multiplied by the quality factor of the radiation.

Solid-state dosimetry: Two methods available for the measurement of integrated dose in natural materials. They are thermoluminescence and electron paramagnetic resonance (EPR); the former is used with ceramic materials and quartz and the latter with tooth enamel. Approximate minimum detectable absorbed dose levels are 10 mGy (1 rad) and 100 mGy (10 rad) for thermoluminescence and EPR, respectively.

Source term: The amount of radionuclides or chemicals released from a site to the environment over a specific period for use in dose reconstruction.

Statistical power: See *power*.

Statistical significance: A finding that, according to specific assumptions and based on mathematical probability, is not likely to have been the result of chance. In epidemiology for example, significance testing is a measure of whether a difference observed between the exposed and nonexposed groups in a study is real or merely a random variation. The probability of an observed difference being the result of chance can be expressed as a p-value.

Stem cell: A cell that can differentiate into any one of several types.

Susceptibility: The sensitivity of different people to the genotoxic effects of ionizing radiation, provided by different amounts of DNA repair capacity or different metabolic levels of biochemicals that prevent radiation genotoxicity.

Thermoluminescence: One of two principal methods of solid-state dosimetry for the measurement of integrated dose in natural materials.

Time dependence of dose: Dependence of the frequency of stochastic effects of radiation from dose rate and latent period make it necessary to take into account time-dependence of all kinds of doses delivered to the whole body or organs. In analyses of epidemiologic data, time-dependence of dose has to be taken into account for effects specific to different age and sex groups.

Transfer factors from environmental data to dose: External dose for different groups of the population can be evaluated from dose rate measurements in open areas or from radioactive contamination as determined by spectrometric methods or by calculation. Internal dose from inhalation and ingestion can be evaluated from radionuclide concentration in air and in food products.

Transuranic: An element with an atomic number greater than that of uranium (92).

Uncertainty analysis: Quantification of the extent of uncertainty in the model result that is due to all conceivable sources. Most commonly, uncertainty analysis involves the probabilistic propagation of uncertainty in the parameters and in the functional terms of a model to provide a probabilistic statement for the model result from which a confidence interval can be obtained for decision-making. This confidence interval is most properly referred to as a "subjective confidence interval" or "credibility interval" given that judgment must be used to quantify the present state of knowledge about components of the model using incomplete or partially relevant data sets.

Vadose zone: The unsaturated (shallow) soil layer that constitutes the region above the level of the permanent groundwater.

Validity: The absence of systematic error or bias in, for example, a set of measurements. In epidemiology, validity also can refer to the degree to which study results are extrapolated to populations other than those in the study sample.

Index